METRIC UNITS IN ENGINEERING
—Going SI

METRIC UNITS IN ENGINEERING —Going SI

How to use the new international system of measurement units (SI) to solve standard engineering problems

Cornelius Wandmacher, P.E.

Professor of Engineering Education
University of Cincinnati

Fellow, ASCE and ICE; Director, ANMC;
Past President, ASEE

INDUSTRIAL PRESS INC.
200 Madison Ave., New York, N.Y. 10016

Dedication

To my family, my colleagues, and many friends whose understanding and encouragement made "Going SI" a most enlightening, rewarding, and memorable experience.

Library of Congress Cataloging in Publication Data

Wandmacher, Cornelius, 1911–
  Metric units in engineering—going SI.
  Includes index.
  1. Metric system. I. Title. II. Title: International system of measurement units (SI) to solve standard engineering problems.
  QC91.W36   389'.152'02462   77-17935
  ISBN 0-8311-1121-6

METRIC UNITS IN ENGINEERING—

Going SI

Copyright © 1978 by Industrial Press Inc., New York, N.Y. Printed in the United States of America. All rights reserved. This book or parts thereof may not be reproduced in any form without permission of the publishers.

Contents

Chapter		Page
	Preface	vi
1	Introduction to SI	1
2	Statics	30
3	Dynamics	57
4	Strength of Materials	88
5	Mechanics of Machines	119
6	Fluid Mechanics	141
7	Thermodynamics and Heat Transfer	190
8	Electricity, Magnetism, and Light	225
9	Conversion to Preferred SI Usage	233
10	Moving Into the World of SI	265
	Appendix I	289
	Appendix II	293
	Index	299

Preface

Intended for the use of practicing engineers and students who are already knowledgeable in the area of engineering principles, this book is meant to be of assistance in applying such knowledge to the solution of typical problems but stated in terms of SI units.

Therefore it is anticipated that the reader will understand and observe in each instance the limitations, if any, of the engineering principles and procedures involved. Examples presented refer, in general, to common or average conditions; the objective in particular is to illustrate, in the most direct manner possible, the essential points to be observed in the utilization of SI.

Since this book treats engineering principles broadly, it is *not* designed to supplant standard texts or references on specific topics which should, of course, be consulted for further interpretation of theoretical background as needed.

A key factor in the successful use of this book, however, will be found in the emphasis on unit check-outs at intermediate and final solution points in the computations. In addition to assuring consistency in the usage of SI, the unit check-out process will rapidly increase the reader's comprehension of SI as a highly desirable and coherent system for engineering work. As will be noted, top priority is given to the demonstration of SI as a new measurement *system*.

Thus, the following outline of problem solving procedure is recommended:

(1) Select an applicable equation for conditions presented, and write it down, using appropriate physical *quantity* symbols;
(2) To the right of this equation make a separate unit check-out using appropriate SI *unit* symbols;
(3) Substitute numbers, in SI terms, for the given magnitudes of physical quantities in the equation;

PREFACE vii

(4) Solve for a numerical answer, keeping in mind the matter of significant digits;
(5) Check the units of the answer and state the answer in terms of preferred SI units or unit multiples;
(6) Where possible, make a general *comparison* of the answer with previous experience based on Customary Units;
(7) Look for and note new personal "bench marks" in SI terms.

A word of caution is offered to those who may wish to compose additional Examples. Do *not* take data in rounded magnitudes of Customary Units and simply convert directly to SI magnitudes. Such procedure usually puts undue emphasis on conversion factors and on unwarranted decimal places. The best examples will be those stated from the outset in rounded magnitudes of SI units. This puts the emphasis where it should be: on the *changeover* to SI as a totally new system.

This book is not intended to be an authoritative source of conversion factors. Where stated, such factors are given for the purposes of comparison and illustration. Although every possible precaution has been taken to assure the correctness of the conversion factors quoted, the latest official sources such as NBS and ASTM should always be consulted when an actual application is to be made.

In case of any inconsistency between the book's illustrative text and the conversion factors quoted from ASTM E 380-76, the latter should, of course, be considered as controlling.

Conversion factors are generally stated to a sufficient number of decimal places to cover a wide range of precision but the actual use of such factors must take into account the question of significant digits applicable to a particular problem.

The starting point for discussion of significant digits is in the statement of original data. After the decision is made as to the preciseness with which data shall be stated, the procedures for recognizing significant digits as outlined herein shall apply in all further computations.

Every engineering project requires that initial decisions be made about the degree of precision of measurements necessary in various parts of the project. This fact leads inevitably to the discussion of the vital area of tolerances, fits, true positioning, clearances, etc. Important as they are, however, these subjects do *not* come within the scope of this book.

The style used in this book is intended to be consistent throughout, with the Federal Register Notice of October 26, 1977, as reproduced in

the NBS Letter Circular, LC 1078, "The Metric System of Measurement (SI)," and with NBS Special Publication 330 (1977 Edition) *The International System of Units (SI)*.
This book reflects a combination of views based on both U.K. and U.S.A. engineering practice and upon experience with metrication utilizing SI in several countries. All possible effort has been applied to achieve the optimum in nomenclature, symbols, and preferred as well as correct usage of SI, and the author assumes full responsibility for all of the content.

ACKNOWLEDGMENTS

Much of the inspiration for this book came initially from the "National Metric Study Conference" sponsored by the Engineering Foundation at Deerfield, Massachusetts, in August 1970. Acknowledgment is due a host of friends and colleagues in ANMC, ANSI, ASTM, ASCE, ASME, ASEE and at the University of Cincinnati who have all made many significant contributions and suggestions since that time.

Primary acknowledgment should go to Louis F. Polk, Sheffield Division of the Bendix Corporation; Roy P. Trowbridge, General Motors Corporation; and John D. Graham, International Harvester Corporation for their initial orientation of the author in the merits and opportunities of "Going SI," and their continuing generous counsel.

Important contributions particularly in the formulation of the early drafts and concepts of this work, were made by the late Alan J. Ede, University of Aston, Birmingham, England, and by Graham Garratt, formerly of Industrial Press Inc.

Advice and information received from abroad was of invaluable assistance, especially that provided by Alan F. A. Harper and Hans J. Milton, Metric Conversion Board of Australia; W. I. Stewart, Standards Association of Australia; Ian D. Stevenson, Metric Advisory Board of New Zealand; Gerald H. Edwards, Standards Association of New Zealand; Stephen M. Gossage and Paul C. Boire, Canadian Metric Commission; P. J. L. Homan, U.K. Metrication Board; H. L. Prekel, Metrication Department, South Africa Bureau of Standards.

The author is especially indebted to (a) colleagues of the ANMC Metric Practice Committee: Russell Hastings, Clark Equipment Co.; Louis E. Barbrow, National Bureau of Standards; Bruce B. Barrow, Institute of Electrical and Electronics Engineers; William G. McLean, Lafayette College; (b) colleagues of the ASCE Committee on Metrication, Andrew Lally, American Institute of Steel Construction; R. Ernest

PREFACE

Leffel, Camp Dresser & McKee; Carter H. Harrison, Jr., Stevens, Thompson & Runyan; (c) colleagues of the ANMC Construction Industries Coordinating Committee: T. Clark Tufts, Tufts & Wenzel, and Anna M. Halpin, McGraw-Hill Information Systems Company, Sweet's Division; (d) colleagues of the ASEE Metric Committee: Harold L. Taylor, Inland Steel Corporation; William Lichty, General Motors Institute; (e) Robert D. Stiehler and Jeffrey V. Odom of NBS; Melvin R. Green of ASME; Malcolm E. O'Hagan of ANMC; Donald L. Peyton of ANSI; and (f) Ronald M. Huston, Louis M. Laushey, William H. Middendorf, and Daniel J. Schleef of the Faculty of Engineering at Cincinnati.

In the final analysis much praise is due to Hans K. Moltrecht, Clara F. Zwiebel, and Paul B. Schubert of Industrial Press Inc.; Peter A. Oelgoetz, draftsman; and especially to Marie Benton of the University of Cincinnati, all of whom worked imaginatively and steadfastly through numerous versions of the manuscript and the illustrations.

All of the experiences in preparing this book have led to an unmistakable conclusion that "Going SI" is truly a highly significant advancement in measurement unit systems, a look to the future, a real "move ahead."

1

Introduction to SI

1.0 A One-World View

Global communications, as well as the transportation of goods and people, have now developed to such an intensity that from an economic point of view, worldwide everyday use of an international system of measuring units is increasingly advisable, and from a technical point of view, is highly desirable as well.

Great impetus to this development was given in 1954 when the principal international governing body in such matters, known as CGPM, the General Conference on Weights and Measures, adopted a much improved *"International System of Units"* for which since 1960, the official abbreviation in all languages is SI.

SI is a *modernized* metric system. The advantages of this new system, notably its simplicity and universality, coupled with its close relationship to the original metric system, led to its almost immediate acceptance on a broad basis. The new system attracted much favorable response particularly from imaginative engineers, scientists, and educators throughout the world and SI is now rapidly displacing the original metric system. All of the nations formerly on the "English" system are now in the process of change to SI. Any country which now is said to be in the process of going metric, is in fact adopting the new International System and is thus "Going SI."

In summary, the development of SI is a major evolutionary step, an advancement for the entire world in developing a much improved measurement system for the use of all nations.

Both the traditionally metric world and that part of the world which formerly utilized the so-called "English" system, will be moving to the improved international system, SI, as shown in Fig. 1-1.

It is true that for the former English-system countries this will mean a new experience in utilizing the meter and the kilogram; but as will be seen, there is much more to SI than m and kg. In fact, those countries which change over from the English system to SI (if the transition is

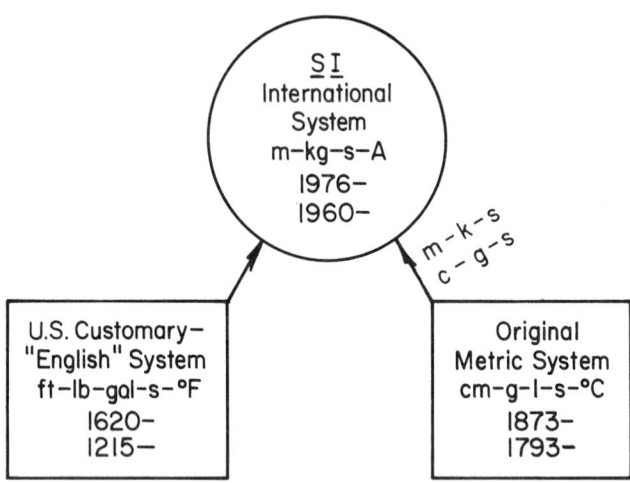

Fig. 1-1. "Going SI"—The Move Ahead.

made wisely and expeditiously), may be reaping many of the benefits of SI far in advance of some of the traditionally metric countries where an air of indifference may well prevail. Going SI is a step ahead for all the world. Those with vision, who make this advance to SI, will be rewarded with many early benefits and the greatest long-time gains.

1.1 Why Go SI?

In addition to various economic reasons for the change to SI, a major advantage will be noted in the following rationale:

a) The English language is now the most generally accepted technical language in the world. It seems desirable to strengthen its position as the principal universal medium of technological communication.
b) To optimize the use of English as the key international technical language, it would be most advantageous to have all English-speaking people utilizing SI as the basis of measurements and standards.
c) Quantitative thinking involving measurements and standards is becoming an increasingly influential aspect of daily life for all people, not only in business and industry but also in the home and in community affairs.
d) A simple, coherent, rational measurement system such as SI will provide the greatest long-range serviceability to the professions and to the public at large.

e) SI is a dynamic system and in some ways is still in the process of maturation. The best way to adjust any anomalies which may now exist is to participate directly in the further utilization and development of SI.
f) Many products and processes involve phenomena from several branches of science; that is, they represent truly interdisciplinary endeavors. It is of real value to have one consistent system that embraces all science rather than separate systems developed by scientists working in limited areas.

Occasionally the suggestion is made that an aggressive campaign should be made to "sell" and to expand the use of the foot-pound-gallon-Fahrenheit system. Such a proposal runs counter to the fact that practical electrical units: volt, ampere, etc., have always been SI units and cannot fit into a foot-pound system. Most countries of the world, including all of the major industrial powers, are now on SI or committed to SI. Every country which has adopted an improved measuring system during the past century has moved to a metric system.

1.2 Impact of SI on Engineering

Worldwide adoption of SI is having two major impacts on engineering practice:

1. *Language*. SI provides a much-improved, universal, precise, and simplified technical language for the *description* of engineering quantities. As such, SI is entering into all aspects of engineering measurements, computations, specifications, and graphics. Forward-looking practicing engineers are becoming conversant in SI units and are learning to utilize this new language effectively. The purpose of this volume is primarily to be of assistance to those engineers who wish to work in SI.

It is important to be clear at the outset that the introduction of SI units does not make any difference in the theoretical basis of engineering calculations. A system of units is no more than an extension of language. To use an analogy, the basic description of a game of tennis remains essentially the same whether the commentator is speaking in English or in French. In the same way, physical laws and physical objects are not dependent on the language used to describe them. The effect of introducing a new unit system is largely confined to the changed appearance of the expressions and formulas used. Nevertheless, some of

INTRODUCTION TO SI

KEY ORGANIZATIONS

Relative to the Conceptualization and Utilization of:

SI—INTERNATIONAL SYSTEM of MEASURING UNITS
(Le Système International d'Unités)

Organs of the Metre Convention

BIPM —International Bureau of Weights and Measures
(Bureau International des Poids et Mesures at Sèvres, France)

CGPM —General Conference on Weights and Measures
(Conférence Générale des Poids et Mesures)
(meets at least once every 6 years)

CIPM —International Committee for Weights and Measures
(Committee International des Poids et Mesures)
(meets at least once every 2 years)

CCU —Consultative Committee for Units

Standards Organizations

 ISO —International Organization for Standardization

 IEC —International Electrotechnical Commission

 NBS —National Bureau of Standards (U.S.A.)

Fig. 1-2. Key Organizations (*Continued next page*).

Ch. 1   INTRODUCTION TO SI   5

 ASTM —American Society for Testing and Materials

 IEEE —Institute of Electrical and Electronics Engineers

 ANSI —American National Standards Institute

 ANMC —American National Metric Council

Fig. 1-2. Key Organizations.

these may be so different in their new forms as to require the development of a new frame of reference.

A few of the changes are so obvious they scarcely need mentioning. SI units differ in magnitude from the corresponding customary units, so that the numbers in familiar formulas are altered. The names are different, and this has wider effects than might be immediately apparent. Some engineers are apt to use unit names rather casually and, for example, to talk about the "horsepower" of an engine when meaning to indicate how big it is, merely because its power output would normally be expressed in terms of the customary horsepower unit. When this is no longer an acceptable unit a more specific term will have to be used.

"Mileage" represents another case in point. Such a word with a root in customary units will become obsolete. More specific terms such as "distance," "fuel consumption," or "service" should be substituted—depending on the characteristic actually being described.

2. *Standards*. SI is having an important impact also on engineering

standards which define policy, rules, techniques, sizes, shapes, modules, etc. There is a new set of international standards emerging which is described, or will be described, in terms of SI units.

New engineering standards are constantly under discussion by such agencies as the International Organization for Standardization (ISO), International Electrotechnical Commission (IEC), the Pan American Standards Commission, and the Pacific Area Standards Congress. In these discussions the United States is generally represented by the American National Standards Institute (ANSI).

For the full titles and official initials of various standards organizations referred to in this book, see Fig. 1-2. Many of the new international standards, although described in SI terms, will be based on present standards; others will be improved standards written in SI. Setting the stage for preparation and use of engineering standards in SI terms is a closely related objective of this volume.

1.3 Merits of SI

As with the understanding of any new instrument or tool, the best engineering approach to becoming conversant and dexterous in SI is to obtain an insight into the key aspects of the overall "system." For this purpose it is well to identify the principal objectives of SI which include:

1. *Simplicity*—the identification of a small number of essential Base Units which may be expanded directly into various sets of Derived Units;
2. *Coherence*—direct one-to-one relationships between Base Units and Derived Units without the use of any intervening multipliers;
3. *Uniqueness of Units*—no duplication of Derived Units; i.e., each Derived Unit is used in the same form and with the same name and symbol in all branches of technology to which it is applicable;
4. *Symbolization*—unmistakable identification of units and of unit multipliers (prefixes) by standard symbols;
5. *Decimalization*—simplified computation and recording similar to the decimal monetary system; utilization of the concept of powers of 10;
6. *Versatility*—units applicable to various requirements with convenient unit multiples and sub-multiples to cover a wide range of sizes;
7. *Universality*—worldwide language in respect to terminology;
8. *Reproducibility*—description of Base Units in terms of reproducible physical phenomena with no dependence on artifacts except for the unit of mass.

Ch. 1 INTRODUCTION TO SI 7

Table 1-1. Base Units

Quantity	Unit Name	Unit Symbol
length	meter	m
mass	kilogram	kg
time	second	s
electric current	ampere	A
thermodynamic temperature	kelvin	K
amount of substance	mole	mol
luminous intensity	candela	cd

These highly desirable and advantageous characteristics of SI will be borne out by a few further comments on each of the significant points:

I—*Simplicity*. SI is built entirely upon seven Base Units, the names and symbols for which are shown in Table 1-1.

II—*Coherence*. A homogeneous family of Derived Units is obtained from the Base Units in a direct one-to-one relationship of these units. Changeover to SI will unclutter one's mind of many multipliers such as: 12, 144, 1728; 3, 9, 27; 4, 16, 32; 7.48, 62.4; 550, 746, 33,000; 5,280; 43,560; etc. *All will be gone!*

An excellent example of the simplifications resulting from a coherent system is shown in the following set of SI Derived Units:

Table 1-2. Derived Units

Quantity	Unit Name*	Unit Symbol	Expressed in Terms of SI Units
velocity	...	...	m/s
acceleration	...	...	m/s^2
force	newton	N	kg·m/s^2
pressure, stress, modulus	pascal	Pa	N/m^2
energy, work, quantity of heat	joule	J	N·m
power	watt	W	J/s

* *Note* that some, but by no means all, of the Derived Units have been given special names.

III—*Uniqueness of Units*. Since there are no alternative or duplicate units for any quantity in SI, all forms of "energy" whether potential, kinetic, mechanical, electrical, or thermal will be measured in terms of "joules."

The direct connection between electrical and mechanical energy is made possible by the new definition of the ampere (A) and its identification as a Base Unit of SI. The SI definition is built upon the concept of *force* developed between two parallel electrical conductors under specified conditions. See Chapter 8, "Electricity, Magnetism, and Light."

Charts of SI. Where graphical interpretations are favored it may be advantageous to show in chart form how the most frequently used Base Units of SI are formed into some key Derived Units. The study of such charts, as given in Chapter 9, "Conversion to Preferred SI Usage," can be highly effective in comprehending both the coherence and uniqueness aspects of units of SI.

IV—*Symbolization.* SI represents a number of advances in the use of symbols for units as well as symbols for prefixes used to denote multiples and sub-multiples of units. The SI symbols are to be used in the same form in all languages of the world.

There are specific rules by CGPM on such matters as type style, capitalization, and punctuation, as follows:

> Upright type, in general lower case, is used for symbols of units; if however, the symbols are derived from proper names, capital type is used (for the first letter). These symbols are *not* followed by a full stop (period).*

Thus m, kg, s are lowercase, but N, Pa, J, and W are capitalized as symbols, although the "unit name" (newton, pascal, joule, watt) is *not* capitalized.

Appropos the matter of symbols it is well at this point also to observe that, in accordance with U.S. and international standards, the symbols for all quantities (length, mass, time, etc.) are indicated by italic type (L, m, t, etc.).

Since more than usual discussion is given over to units and unit checkouts in the text and examples which follow, it is essential at all times to observe the distinctions of upright type for "units" and italic type for "quantities."

Thus m denotes "meter," but m denotes mass. For further information on this subject, see Chapter 9.

SI symbols are *not* abbreviations and therefore are *not* to be treated as such. They are to be used, each in its own right, as a "symbol." In view of the greatly increased emphasis on symbols in SI, it is imperative that each SI symbol be used exactly in its approved standard form.

* From National Bureau of Standards, NBS 330.

Standard SI "prefixes" build upon an ingenious idea of the original metric system which used prefix symbols to denote levels of magnitude of unit quantities, with increments related to powers of 10. An augmented set of prefixes is used in SI to form multiples and sub-multiples of all Base Units and of Derived Units, thus greatly reducing the effort of memorizing many special names.

Rules similar to those concerning capitalization and omission of punctuation in unit symbols also apply to prefix symbols (see examples that follow). Note that mega, giga, tera, (M, G, T) are capitalized in symbol form in order to avoid confusion with already established unit symbols, but that they maintain the lowercase letters when the prefix name is spelled out.

V—*Decimalization*. The elimination of numerical fractions and accompanying compound arithmetic, as accomplished in the original metric system, is carried several steps further and strengthened in SI. Decimal arithmetic is facilitated not only by the introduction of additional prefixes but also by the expressed preference for intervals of 3 in the "powers" of multiples and sub-multiples. The prefixes generally preferred in engineering are shown in Table 1-3.

Note that as with any decimal notation there is a particular need always to be mindful of the number of "*significant digits*" in the outcome of any computation. (This matter is discussed further in Sec. 1.7 and in Chapter 9.)

VI—*Versatility*. The Base Units should be such as to be readily comprehensible and to be usable directly in numerous everyday instances. Thus the meter, the kilogram, the second, and the ampere all represent reasonably familiar quantities. The kilometer (km) and millimeter (mm) can also be readily visualized and are found to be generally useful. It will be noted that the "kilogram," with a prefix title, may seem to be named inconsistently as a Base Unit under SI principles.

Table 1-3. Most Frequently Used Prefixes

* Multiples			* Sub-Multiples		
Factor	Prefix	Symbol	Factor	Prefix	Symbol
10^9	giga	G	10^{-3}	milli	m
10^6	mega	M	10^{-6}	micro	μ**
10^3	kilo	k	10^{-9}	nano	n

* A complete set of prefixes is given in Table 10-1.
** The upright (roman) type style is preferred when available, as with all other prefixes.

However, the kilogram for some time has been a widely accepted "unit" throughout the world. It is the only Base Unit in SI still dependent on an artifact. As such it was considered appropriately sized and well-identified, thus the decision was made to recognize and define the kilogram as the Base Unit of mass.

VII—*Universality*. A strong endeavor has been made in SI to achieve universally acceptable titles for units. Obvious emphasis has been placed on utilizing the names of widely regarded applied scientists (Newton, Pascal, Joule, Watt) in order to eliminate or minimize language translation difficulties.

In order to eliminate an instance of confusion in a title such as "centigrade," for example, which linguistically impinged on another unit name, "centigrad," commonly used in some countries for angular measure, the former name of the ambient temperature interval was changed to "degree Celsius" in honor of the Swedish astronomer who introduced the 1/100 temperature scale.

The SI-related symbol is continued as: °C. Note especially that in this instance the use of the degree symbol (°) avoids conflict with the symbol for coulomb (C).

VIII—*Reproducibility*. For any worldwide system, reproducibility (or comparability) of references for Base Units is of prime importance.

Hence, SI defines all possible Base Units (except the kilogram) in terms of scientifically reproducible physical phenomena. While it is seldom necessary in engineering activity, SI units can now be reproduced, as may be desired, by competent scientific personnel in various physical laboratories around the world.

1.4 Changeover versus Conversion

There is a vital reason, in an engineering sense, to observe and to encourage the following distinctions in terminology that will be utilized in this book:

- a. *"Changeover"* is best used to denote an overall change from the use of the U.S. Customary System to the use of SI as a complete framework of reference. Thus, the entire process of "Going SI" is most properly referred to as a *system* changeover.
- b. *"Conversion"* generally refers to the determination of equal numerical values of quantities and units by means of tables, charts, diagrams, and arithmetic methods.

Conversion, therefore, deals with the limited and temporary arithmetical aspects of "Going SI." It can become a "numbers game" in which the participants all too often lose sight of the real objective. For the most fruitful experience with SI it is vital not to get bogged down with excessive emphasis on numerical conversion routines. Accordingly, this book will deal chiefly with the system concepts of changeover.

1.5 Preferred Usage of SI

Since simplification and improved communication in measurements are objectives of "Going SI," and since SI is designed as a coherent, comprehensive unit system, it follows naturally that it is advisable to make decisions about certain "preferred usage."

Decisions about physical measurements are made by BIPM (see Fig. 1-2), but recommendations about SI unit development are made by CCU to CIPM and then to CGPM. The identification of international standards is handled through ISO and IEC. Authority for official interpretation or modification of SI for use in the U.S.A. resides in NBS. The principal working reference document is: "American National Standard Metric Practice," ANSI Z 210.1-1976 (successor to ASTM E 380-1974 and IEEE Std. 268-1973).

Although a wide range of multiples and sub-multiples is provided in SI, only a few are likely to be selected and used in any particular field of technology. The actual choice of the appropriate multiple of an SI unit is governed by convenience, the multiple chosen for a particular application being the one which will lead to numerical values within a generally accepted practical range. For instance, numerical values of 1 to 999 are usually favored; this is consistent with the recommendation for 10^3 intervals in prefixes. In addition, there are certain units outside of SI which are nevertheless recognized by CIPM as having to be retained and to be used in conjunction with SI either because of their practical importance or because of their use in specialized fields. All of these decisions are shown and discussed in ANSI Z210.1-1976 of which ASTM and IEEE are joint sponsors.

1.6 Guidelines for Use of SI

There are some particular characteristics of SI which merit further discussion especially in respect to: (a) length, area, volume; (b) mass, weight, force; and (c) pressure, stress, work, and energy.

1. *Length or Distance.* The usual expression of length, distance, or

dimension in engineering work preferably should be stated in terms of one of the following: km, m, mm, μm. It is generally recommended that the centimeter (cm), commonly used in the original metric system, be de-emphasized in preferred SI practice and used only sparingly. A major reason for this is to give fewer reference points to which to relate a new sense of proportion. In other words, it is believed more desirable to develop a new sense of metric proportion using km, m, mm, or μm wherever possible.

An outgrowth of this point of view, for instance, would be to make construction site drawings mainly in terms of meters and decimals of meters only, while drawings of structural components and machinery parts might be made principally in terms of millimeters only. On many drawings and sketches it will be found possible to include one convenient footnote to state: "All dimensions in meters," or, "All dimensions in millimeters."

A corollary to this is the fact that m and mm being separated 1000-fold are not likely to be confused, while cm and mm being separated only ten-fold, may occasionally be confused.

In some instances such as in everyday layman's affairs, in wearing apparel (body sizes), and in agriculture, where the precision of the mm is not warranted, the "cm" is likely to continue to prevail. But most meter "sticks"—rulers, drafting scales, and measuring tapes—can be expected henceforth to be graduated and numbered in terms of m and mm, rather than in cm. Chiefly, this will mean a change in numbering style; i.e., sub-division marks will be for 10 mm, 20 mm, etc.; thus the centimeter will be viewed as ten millimeters (1 cm = 10 mm).

2. *Area*. The unit of area is, of course, the square meter, m^2. In terms of millimeters this becomes 1 000 000 mm^2, or 10^6 mm^2. Most cross-sectional areas of machinery and structural parts will be in terms of mm^2.

In respect to large land areas, the preferred multiple would be the square kilometer (km^2) which in terms of meters is 1 000 000 m^2. Another acceptable land unit, although *not* an ISO preference, is the "hectare" (ha = hm^2) which is 10^4 m^2, and is approximately 2.5 acres. The hectare probably will be continued in use for some time, but should be recognized as an expedient departure from the 10^3, 10^6, 10^9 sequence of multiples usually preferred. SI prefixes, such as m and k, should never be applied to hectare since this unit already includes a prefix.

3. *Volume*. Generally, volume will be expressed in terms of m^3, as in the case of density, which will preferably be stated as kg/m^3.

A significant improvement in the description of capacity is the declaration by CGPM in 1964, that the word "liter" (L), may be used as a special name for the cubic decimeter. This leads, in turn, to the preferred use of the term "milliliter" (mL) and correspondingly deprecates use of the former term "cc."

However, it is preferred that the term "liter" be used only for liquids, gases, and particulates, and that the milliliter (mL) be the only multiple or sub-multiple.

It has taken some time to arrive at a general consensus on a symbol for "liter"; originally it was a lower case "l." To avoid potential confusion in typing of the lower case "l" and the number "1," it has been common practice to write out "liter," or to use the script form in type (ℓ). In June 1976, CCU made a strong recommendation in favor of using the symbol "L." The Department of Commerce, through NBS, has since ruled in favor of the symbol "L" for use in the U.S.A.

CGPM does recommend that the name liter not be employed to give the results of high-accuracy volume measurements.

4. *Mass.* The kilogram is the SI unit of mass. It is equal to the mass of the international prototype made of platinum-iridium and kept at BIPM under conditions specified in 1889 by the first CGPM.

Although the gram was originally conceived as the metric unit of mass for scientific work, it is generally considered to be a size too small for general technical reference. Accordingly, in SI the kilogram has been declared as the Base Unit.

The fact that the name of this "Base Unit" contains a prefix is not viewed as any handicap. The unit is generally referred to in terms of the symbol "kg." However, it must be observed that in the matter of mass, all prefixes apply to the gram as such, thus:

1.0 kg = 1000 g;   and   1.0 Mg = 1 000 000 g

The preferred units of mass are milligram, gram, kilogram, and megagram (or metric ton), which are written respectively:

mg, g, kg, Mg, (or, t)

The metric ton (t) appears to be carried forward at this time as a deference to wide common usage. Actually the metric ton is 1000 kg and hence, really 1.0 Mg, a recognition which probably will generally be achieved in the years ahead. Thus, in SI use of the term "megagram" (Mg) is preferable. The name "tonne" is not recommended.

A valuable SI fact is that for all practical purposes, one cubic meter of fresh water has a mass of one metric ton (1.0 t) and one liter of fresh water has a mass of one kilogram (1.0 kg). See Fig. 10-1.

Machines and devices for ascertaining the "mass" of a body or of a quantity of material should be calibrated in grams, kilograms, or metric tons, and decimal parts thereof. In the changeover to SI in South Africa, the term "massing" is used to replace the common term "weighing," but this terminology has not readily been adopted in other parts of the English-speaking world.

The ordinary weighing machine or "scale" must be viewed as merely "ascertaining mass." Such a device is calibrated to take into account the conditions of gravity prevailing at the place of use. Thus the calibration of the device is made to give a read-out in the desired terms of mass (kilograms).

5. *Weight* is an ambiguous quantity name in customary usage. On one hand it infers mass, as in the customary title "weights and measures." On the other hand it is recognized as the description of the force resulting solely from gravitational attraction. SI offers a new opportunity to clarify the distinction between mass and force by clearly stated application of the Base Unit for mass, kg, and the Derived Unit for force, N.

In technical language, the quantity names, mass and force, must prevail and be used properly. On the other hand, there seems to be little to be gained at this time by insisting that the same distinctions be made immediately by all laymen. Perhaps it is wiser to accept from the

Quantity	Mass, Length, Time (absolute) SI	Mass, Length, Time (absolute) English	Force, Length, Time (gravitational) Metric	Force, Length, Time (gravitational) English	Customary Combined System
Mass	kg	lb	$kp/(m/s^2)$	$lbf/(ft/s^2)$ slug	lb (alt.: lbm)
Force	$kg \cdot m/s^2$ N newton	$lb \cdot ft/s^2$ pdl poundal	kp kilopond (alt.: kgf)	lbf	lbf
Coherence Factor	1	1	1	1	1/32.17

Fig. 1-3. Comparison of Unit Systems.

layman, for the time being, the meaning that is inferred by the unit used; i.e., if "weight" is quoted in kilograms, it is mass that is intended.

The technical manner in which mass and force (and thereby "weight") are handled in various unit systems may be summarized as shown in Fig. 1-3. In this comparison, and all related discussions, the symbol lb denotes mass in the same manner that kg denotes mass; the symbol lbf denotes force in the same manner as kp (sometimes shown as kgf) denotes force.

From Fig. 1-3 it will be observed that the mass and force quantities in the several systems are related by the following factors:

Based on customary gravitational usage:

Mass: 1.0 slug ≈ 32.17 lb ≈ 14.59 kg ≈ 1.488 kp/(m/s^2)
Force: 1.0 lbf ≈ 32.17 pdl ≈ 4.448 N ≈ 0.4536 kp, (alt.: kgf)

Based on SI usage:

Mass: 1.0 kg ≈ 2.205 lb ≈ 0.06852 slugs ≈ 0.1020 kp/(m/s^2)
Force: 1.0 N ≈ 7.233 pdl ≈ 0.2248 lbf ≈ 0.1020 kp, (alt.: kgf)

The *force* definitions are:

The "newton" is the force required to accelerate one kilogram mass at the rate of 1.0 m/s^2;

The "poundal" is the force required to accelerate one pound of mass at the rate of 1.0 ft/s^2;

The "kilopond" (sometimes called kilogram-force) is the force required to accelerate one kilogram mass at 9.80665 m/s^2;

The "pound-force" is the force required to accelerate one pound of mass at the rate of 32.1740 ft/s^2.

The related definitions for the *derived mass* units are:

The "slug" is that mass which when acted upon by one pound force will be accelerated at the rate of 1.0 ft/s^2.

The gravitational metric unit of mass, is that unit of mass which, if acted upon by a kilopond of force, will be accelerated at 1.0 m/s^2. There seems to be no generally accepted name or symbol for this gravitational unit of mass, except the inference to the kilogram.

The "kilogram" and the "pound" are Base Units, *not* Derived Units as are the slug and the gravitational metric unit of mass. Both the kilogram and the pound relate directly to an artifact of mass which, by convention, is regarded as dimensionally independent, thus the title, Base Unit.

6. *Force*. All of the foregoing is obviously in accordance with the fundamental law of physics ($F \propto ma$), which states that all force is dependent solely on mass and on acceleration. Since SI is a coherent system:

1.0 kg accelerated at 1.0 m/s^2 ≈ 1.0 unit of force ≈ 1.0 newton ≈ 1.0 N

The use of the unit title "newton" should fix in the mind of the user the full significance of the distinctions between mass and force.

Normally, a mass to be supported or moved will be specified or labeled in kg, but all forces acting upon a structure or a machine part, either gravitationally or laterally (including wind, sway, impact, etc.), should be specified or determined ultimately in terms of newtons (N). Distributed forces should be stated in terms of N/m or N/m^2, as the case may be. The most useful multiples will probably be kN and MN; occasionally the sub-multiple mN will be applicable.

Correspondingly, forces or moments applied by testing machines, dynamometers, or similar devices should be specified, recorded, and/or determined in terms of N, or N·m, etc. Deflection calibrations will generally be in terms of mm, or μm, in radians, or in decimal degrees.

If a mass is likely to be hoisted at accelerations of substantial magnitude then, of course, it is essential to determine and state the rating of the equipment in terms of newtons. A graphical representation of the key relationships between mass, force, and weight is given in Fig. 1-4. These relationships will be discussed further in Chapter 2, "Statics," under the topic of Loads and also in Chapter 3, "Dynamics."

The standard (or average) acceleration of gravity at the surface of the earth in SI terms is generally taken in engineering as 9.8 m/s^2. From this it follows that the SI gravitational force, customarily called "weight," resulting from one kilogram of mass being subjected to standard gravitational acceleration, is 9.8 N.

Mass is the measure of a body's resistance to acceleration. Because this property is independent of the body's location, whereas gravitational force is not, a system based on mass is called "absolute" to distinguish it from a system based on force of gravity (which is called "gravitational"). An absolute system becomes more desirable as engineering considerations increasingly involve dynamic conditions in which "mass" is the key quantity.

Abroad, especially in European countries, there has been a custom in engineering to function on a metric gravitational system by introducing the term "kilopond" for which the symbol has been kp. This is alternately called the kilogram-force (kgf). In making the changeover to

Ch. 1 INTRODUCTION TO SI 17

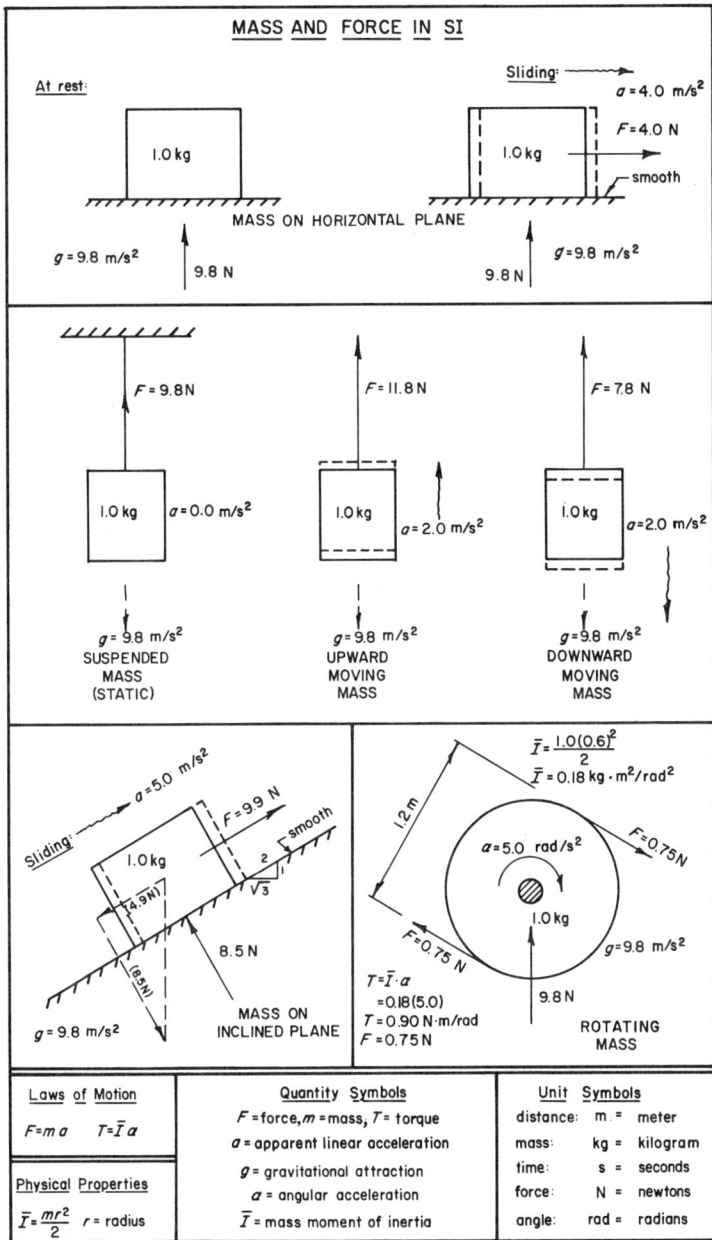

Fig. 1-4. Mass and Force in SI.

SI it has been deemed preferable, as a progressive step, to move directly to an absolute system.

The present use abroad of kp (and kgf) is on the wane. In fact, a recent decision by the European Economic Community is designed to remove the legal status of kp (and kgf) in the near future, thus making it unacceptable in EEC specifications and standards thereafter. Accordingly, this is not the time to expand the use of this term.

7. *Moment.* (Bending and Torsion) The concepts of bending moment and torsional moment both involve the product of a force and a distance, the latter being termed the moment arm. Thus the SI unit product is the newton-meter, which may be symbolized as N·m, kN·m, etc.

Note the word order above, giving first emphasis to the force in newtons. This is recommended to indicate that the symbol should be written N·m, and *not* mN, to eliminate any chance of confusion with a prefix notation which would be read as millinewtons.

Also, it may be noted that moment N·m is a product of a force and a distance *perpendicular* to its line of action. Hence this unit product of static moment should *never* in itself be further interpreted into "joules" which pertain only to work and energy considerations. Nor should torque be stated in joules unless rotation occurs, in which case the units would be newton-meters per radian (N·m/rad). See Chapter 3, "Dynamics."

If for some reason it is desired to revert to Base Units, as the full form of the newton is kg·m/s^2, the newton-meter (N·m) may be expressed also as: kg·m^2/s^2.

8. *Pressure, Stress, Elastic Modulus* may all be stated directly either in N/m^2; or in pascals, Pa. Common multiples are kN/m^2, MN/m^2, GN/m^2; or kPa, MPa, GPa. In appraising the use of the "pascal" it is essential to observe that this important new SI unit was adopted by CGPM as recently as 1971.

It should be noted that MN/m^2 is equal to N/mm^2. This latter form is sometimes used for stress. However, it is generally preferred practice to use the prefix in the numerator only, hence the forms MN/m^2 and MPa are recommended.

Another form of pressure measurement is the "bar" which is 10^5 Pa. This has in the past been favored in some scientific and engineering activities, such as meteorology, fluid power and thermodynamics, since it approximates one atmosphere. As will be noted, the bar is 100 kPa and

0.1 MPa, which is a departure from the 10^3, 10^6, 10^9 intervals usually preferred. The millibar, 10^2 Pa, is a further aberration of preferred usage. Since the use of the bar does not appear to be expanding in other metric countries, there seems to be little point in encouraging any additional use. The kPa and MPa are much preferred wherever SI is being introduced.

For a further discussion of pressure, see Chapter 6, "Fluid Mechanics."

9. *Moment of Inertia ($\bar{I}$)*. The mass moment of inertia of any body relating to rotation about a given axis is the second moment of the particles of that mass about the given axis and, as such, is given generally in kg·m²/rad². The radius of gyration would normally be given in m/rad. For further discussion see Chapter 3, "Dynamics."

10. *Second Moment (I)* and *Section Modulus (S)* of the cross section of structural sections or machine parts are usually preferred in terms of 10^6 mm⁴ and 10^3 mm³, respectively, to be consistent with other dimensions of sections which will usually be given in mm.

11. *Angular Measure*. The radian (rad), although not a Base Unit, is specifically identified as a "supplementary unit" and, as such, is the preferred unit for measurement of plane angles. The customary units of degrees, minutes, and seconds of angular measure are considered to be outside of SI, but are acceptable where there is a specific practical reason. If degrees are to be used, a statement of parts of degrees in decimals is preferred. The SI unit of solid angle is the steradian (sr).

In some engineering activities, however, it is most desirable to define relationships by slopes or proportions without the use of units. This practice will, of course, be greatly simplified by change to SI. A slope formerly given in terms of 12, as in 3 on 12, will become, in SI terms, a slope of 0.25, or if desired: 2.5 on 10.

12. *Energy and Power*. By definition, one newton (N) is: "that force which gives to a mass of one kilogram (kg) an acceleration of one meter per second squared (m/s²)."* In turn, one joule (J) is: "the work done when the point of application of a force of one newton (N) moves a distance of one meter (m) in the direction of the force."* And one watt

---

* See C. H. Page and P. Vigoureaux, eds., "The International System of Units (SI)," NBS Special Publication 330, 1974.

(W) is: "the power which in one second (s) gives rise to the energy of one joule (J)."* Specifically in terms of energy:

$$N \cdot m = J = W \cdot s$$

This relationship is one of the most significant advances in Going SI. For further discussion of this point and a more explicit definition of the watt (W), see Chapter 5, "Mechanics of Machines."

Since one joule per second is one watt, conversely one joule is one watt-second. It can be expected that the kilowatt-hour (kWh) will in all likelihood be succeeded by the megajoule (MJ), a move which is already occurring, since

$$1.0 \text{ J} = 1.0 \text{ W} \cdot \text{s} \quad \text{and} \quad 1.0 \text{ kWh} = 3.6 \text{ MJ}$$

SI provides the first opportunity in an energy-conscious world, for a simple, direct measure between mechanical and electrical energy. In the future it will thus be possible to give in watts (W), in kilowatts (kW), or in megawatts (MW), the power ratings of: automobiles and trucks; oil, gas, coal and electric heating systems; refrigerators and air conditioners; lighting, cooking, and communication systems; pumps, engines, and turbines; chemical and nuclear reactors; etc.

13. *Temperature.* Being an absolute and a metric system, SI refers basically to the "kelvin" thermodynamic temperature scale. On this scale either a temperature or a temperature interval is known as a kelvin (symbol K). Although not a part of SI, the Celsius temperature scale for which the zero reference is the freezing point of water, is used with SI. The special name "degree Celsius" (symbol °C) is given to a unit of this scale; it is identical with a kelvin. In effect, the centigrade scale is renamed Celsius and used with SI.

14. *Time.* The SI base unit of time is the second (s) which is preferred in most technical expressions and calculations. It is of interest to note that the second (s) is the only measurement unit which already has complete international acceptance. While there have been proposals for a decimalization of the customary 24-hour day, these have generally been deferred in recognition of the present universal serviceability of the second (s).

Use of the hour (h) as in km/h, and the day (d) as in m^3/d, will occur in special cases, but the use of the minute (min) will be de-emphasized in favor of the second (s). For this reason the statement of speeds of rotating machinery will move toward a preference for revolutions per second (r/s), except in the case of slowly rotating machinery where the r/min may be used in order to keep the quantity in integers.

For high speeds there is already wide use of the millisecond (ms), the microsecond (μs), and the nanosecond (ns). The kilosecond (ks)—which is about one quarter of an hour (actually 16.67 minutes)—has much potential for wider use.

It may be of interest to note that there are 86.4 ks/d (kiloseconds per day) and 31.56 Ms/a (megaseconds per year). Some "old timers" may even wish to observe that a person in his early 60's can be said to be "less than 2.0 Gs of age." In any case it may be of value to observe that:

$$1.0 \text{ ks} \approx 16.67 \text{ min} \qquad 1.0 \text{ Ms} \approx 11.57 \text{ d}$$
$$1.0 \text{ Gs} \approx 11\,570 \text{ d} \approx 31.68 \text{ yrs.}$$

15. *Dates.* While not a part of SI, it is worth noting that ISO has also made a recommendation for standardization in the statement of dates. Under the adopted style, the date of the U.S. Bicentennial was 76-07-04. This recommended form of dating has the obvious advantage of placing all reference to dates in numerical as well as in chronological order.

16. *Abandoned Units.* For various reasons many derived and specialized units will fall into disuse as the changeover to SI progresses. Some, like Btu and horsepower will be dropped because they are based on the U.S. Customary system.

But many former metric units, of the CGS variety, are no longer recommended by CIPM. These include: erg, dyne, poise, stokes, gauss, oersted, maxwell, stilb, and phot. In addition, a number of metric units are outside of SI and their use is deprecated. These include: kilogram-force, calorie, micron, torr, fermi, metric carat, stere, and gamma. The loss of these familiar units may cause a temporary distress in some instances but the long-range total benefits of complete utilization of SI will generally far outweigh the incidental losses in a relatively short time.

1.7 Significant Digits

When numerical conversions are to be made it is especially important to keep in mind the all-too-frequently neglected principles of "significant digits."

In general, the result of any multiplication, division, addition, or subtraction cannot be given in any more significant digits than were given in any one component of the *original data*. This condition pertains regardless of the number of decimal places in which a conversion factor is given.

It is desirable to have conversion factors stated in reference tables to a substantial number of decimal places in order to cover a wide range of uses, but it is the responsibility of the engineer to use the decimal number resulting from a conversion calculation only to the extent that it is applicable.

The following example will briefly illustrate this point and show also that the rule about significant digits works both ways; i.e., SI to customary English units, as well as the reverse, customary English units to SI units.

Consider the question: "What is the equivalent of 3 miles in terms of kilometers?"

Conversion Factor	Direct Multiplication	Significant Equivalent
miles to kilometers = 1.609	3 mi ≈ 4.827 km	3 mi ≈ 5 km

The result is similar in the reverse form:

| kilometers to miles = 0.6214 | 5 km ≈ 3.107 mi | 5 km ≈ 3 mi |

A more complete treatment of this topic is given in Chapter 9, "Conversion to Preferred SI Usage."

1.8 Equivalents of SI Quantities

Some typical engineering examples given in Table 1-4 will illustrate, in general terms, the key principles of handling conversions and the proper determination of significant digits in the equivalent SI magnitude.

An extreme example of a well-intentioned but overzealous endeavor to "Go SI" is the action of a major league baseball team which had the 392-ft sign on the outfield fence repainted to read 119.482 m, and then proudly proclaimed it was now hitting "metric" home runs. More correctly such a sign should read: 119 m.

Another example of inept conversion is the newspaper illustration which goes overboard in precision by announcing that a 5-pound infant

Table 1-4. Equivalents of SI Quantities

Quantity	In SI Terms		U.S. Customary Unit Equivalent	
distance	5.4 kilometers	km	3.4 (or 3 3/8) miles	mi
dimension	2.34 meters	m	7.68 ft (or 7'-8")	
thickness	19 millimeters	mm	3/4 in.	
area	4.00 hectares	ha	9.88 acres	
velocity	6.0 meters/second	m/s	19.7 ft/sec	fps
flow	0.02 meter^3/second*	m^3/s	300 gallons/min	gpm
	0.10 meter^3/second		3.5 cu ft/sec	cfs
	4.0 meter^3/second		91 million gals/day	mgd
pressure	12 meters of water		39 ft of water	
	150 kilonewtons/meter^2	kN/m^2	22 lbs/sq in.	psi
	or 150 kilopascals	kPa		
stress	100 meganewtons/meter^2	MN/m^2	15,000 lbs/sq in.	psi
	or 100 megapascals	MPa		
modulus	200 giganewtons/meter^2	GN/m^2	30,000,000 lbs/sq in.	psi
	or 200 gigapascals	GPa		

* Preferred usage in general text material would require the full word statement, "cubic meters per second", etc., rather than the abbreviated style used in this Table.

had an equivalent "weight" of 2268 grams, whereas 2300 grams is the more appropriate statement.

In the computation of areas and volumes, and in all other instances of arithmetic products, similar caution must be applied. For instance, in determining the area of a floor measured and described as,

$$\text{width} = 3.5 \text{ m}; \quad \text{length} = 62.642 \text{ m}$$

$$3.5(62.642) = 219.2270$$

which indicates a probable area of 220 m^2.

Had the width been stated to a greater degree of precision, then a probable area to more significant digits would result, as follows:

Width (m)	Length (m)	Area (m^2)
3.50	62.642	219
3.500	62.642	219.2

1.9 Notation and Numerical Style

1. *Decimals.* The dot on the line continues as the conventional decimal point marker as in the U.S. Customary (English) System. When the value is less than unity the decimal point should always be preceded by a zero. thus: 0.8, 0.74, 0.062, etc.

Whole numbers do not need a decimal marker as in 12, 640, 3200, etc., unless the intent is to indicate precision to the last given place, such as: 23.0, 640.0, etc.

In some countries the decimal point is put at mid-height of the line of digits. In others, the comma is used for a decimal point. But these styles are not recommended in the United States at the present time.

2. *Grouping of Numbers.* It is recommended that the comma as a thousands marker be omitted. Instead, numbers preferably should be grouped in sets of three as follows:

$$23\,600 \qquad 1\,500\,000 \qquad 2\,676.402\,06$$

An exception is permitted when the number does not exceed four digits on either side of the decimal point in which case the number may then be written:

$$5296 \qquad 0.6012 \qquad 1001.6$$

However, in a column with other numbers that show space and are aligned on the decimal point, the space is necessary.

3. *Spacing Numbers and Units.* One space should be left between any number and the name or symbol of a unit, as indicated:

$$64\,\text{kg} \qquad 20.6\,\text{W} \qquad 250\,\text{milliliters} \qquad 16\,000\,\text{MN}$$

4. *Multiplication Sign.* A centered dot is preferably used as a multiplication sign between unit symbols as indicated below.

$$\text{N} \cdot \text{m} \qquad \text{kg} \cdot \text{m/s}^2 \qquad \text{W/(m} \cdot \text{K)}$$

The symbol "×" is the multiplication sign for numbers: 2.4 × 0.6.

5. *Division Sign.* The oblique stroke, / (slash), a horizontal line to denote a numerator and a denominator, or a negative exponent, may be used to denote the division of one unit or a combination of units by another, for instance:

$$\text{m}^3/\text{s} \qquad \text{or,} \qquad \frac{\text{m}^3}{\text{s}} \qquad \text{or,} \qquad \text{m}^3 \cdot \text{s}^{-1}$$

6. *Value of Prefixes*. The availability of the metric prefixes makes it possible to adjust the size of a unit very simply, without the need for the sort of arithmetic involved in changing from square feet to square inches. On the other hand, there is a need to develop an awareness of the importance of the new prefixes and a facility for handling them. For instance, one must not overlook a "k," thereby introducing a thousand-fold error. However, the value of using prefixes only in incremental powers of three is that a thousand-fold error is not likely to go undetected. Nevertheless, clarity and precision in symbolic writing are just as essential as accurate articulation in reading and speaking.

7. *Prefixes are Shorthand*. As an example, consider a moment or a torque in which the force in newtons and the lever arm in meters are both provided with prefixes. There is nothing wrong with either N·mm or kN·m; nor is there anything wrong with kN·mm, but this is a needlessly long way of writing N·m, since the kilo and the milli cancel out. It is useful, therefore, to recognize that a compound unit such as this, is nevertheless a unit in its own right. If a moment is produced from a force measured in kilonewtons, and a length measured in millimeters, the resulting unit should not be thought of merely as the product of kN and mm, with the two units regarded as existing quite separately. The primary unit for moment is the N·m, and nothing is achieved by arguing about where to put several prefixes that may appear appropriate. A prefix is merely a shorthand way of recording the applicable exponent of ten. Hence the shortest possible notation should be used.

8. *Prefixes Not Spaced*. Words and symbols for prefixes should be written as a part of the unit word or symbol, without spacing, for example:

$$\text{meganewton (MN)} \quad \text{milligram (mg)}$$

9. *Apply Prefixes to Numerator*. It is generally recommended that only one prefix is used, usually in the numerator, to form the multiple of a compound unit:

$$\text{mm/s,} \quad not \quad \mu\text{m/ms} \quad \text{L/m,} \quad not \quad \text{mL/mm}$$

However, since the kilogram is an SI Base Unit, it is proper to write: kJ/kg, etc.

Also, in special cases where the usual measurement unit for the denominator employs a prefix, it may be advisable to standardize on a unit such as N/mm for spring rate.

10. *Prefixes Not Mixed*. Each prefix must be used only in its own

right and *not* mixed with other prefixes; i.e., compound prefixes formed by the juxtaposition of two or more prefixes are *not* to be used, for example:

$$1.0 \text{ nm} \quad not \quad 1.0 \text{ m}\mu\text{m}$$

11. *Units Not Mixed.* A mix of units, as in the U.S. customary English system where "feet and inches," or "pounds and ounces" are given, is not necessary or proper, in SI. Only *one* unit, or multiple of a unit, should be used in writing any quantity, as indicated:

$$20.160 \text{ m} \quad not \quad 20 \text{ m } 160 \text{ mm}$$

$$75.050 \text{ kW} \quad not \quad 75 \text{ kW } 50 \text{ W}$$

12. *Use of Exponents.* When it is desired to express the magnitude of a very large (or very small) quantity in absolute numeric terms, exponential indicators in terms of the power of 10 should be used, such as:

$$3 \times 10^6 \text{ N} \quad 6.2 \times 10^{-5} \text{ Pa} \cdot \text{s} \quad 10^3 \text{ mm}^4$$

13. *Exponent Interpretation.* An exponent attached to a symbol containing a prefix indicates that the multiple or sub-multiple of the unit is raised to the power expressed by the exponent, for example:

$$1.0 \text{ mm}^3 = 1.0(10^{-3} \text{ m})^3 = 10^{-9} \text{ m}^3$$

$$1.0 \text{ mm}^{-1} = 1.0(10^{-3} \text{ m})^{-1} = 10^3 \text{ m}^{-1}$$

1.10 "Soft" versus "Hard" Changeover

At an early stage of moving into SI it is valuable to develop a broad understanding of the available options in both extent and sequence of changeover.

The plan or pattern of movement into SI is generally referred to in terms of a degree of "hardness" which may be described as follows:

a) "Soft" changeover means describing existing things in terms of SI units, but *not* changing sizes, except within tolerances already permitted in U.S. Customary units.

b) "Hard" changeover means designing, describing, producing, and using things entirely in terms of rationalized SI sizes and specifications, including the development of new tolerances in SI.

It is generally desirable to make a "hard" changeover whenever possible and that approach will be utilized in the illustrative problems of this book.

A "soft" changeover may be found most practical in respect to certain series of engineering items or activities for a limited period of time, but it is generally advantageous to aim toward total SI as part of any long-range, progressive plan for complete changeover.

Occasionally, reference is made to a "compatible" changeover in which conformance to SI in the most critical characteristic is accomplished within the product by internal adjustment without major change in design. An example is the volumetric capacity of a soft drink bottle in which the shape is modified slightly to contain 500 mL without changing any critical external dimension which would have an impact on filling, capping, handling, or packaging.

There are, of course, many other examples where changeover to SI is made initially only at a selected, suitable interface of components.

1.11 Rationalization and Optimization

Changeover to SI requires many decisions in respect to the standard sizes of items but at the same time it presents an extraordinary, once-in-a-lifetime opportunity for accomplishing the following:

a) Evaluating the present range of sizes of various parts and products;

b) Identifying new families of parts;

c) Improving the interrelationships between design, production, fabrication, construction, and utilization systems.

In connection with such activities it is also well to establish agreement and common understanding on the following intimately related terms:

Rationalization refers to the establishment of standard SI sizes of materials, components, parts, etc., wherever possible in terms of convenient simple integers.

Optimization generally refers to achieving the minimum number of different stock or standard sizes while at the same time satisfying a wide range of demands for the item. In all discussions of "sizes" it is, of course, essential to distinguish actual sizes from nominal sizes.

1.12 Modules and Preferred Sizes

In connection with achieving the objectives above, two other opportunities will be presented. These concern the determination and/or evaluation of preferred sizes and standard modules.

Preferred Sizes refers to a series of items for which a *geometric progression* is the guiding factor in the determination of capacity intervals, as in: pipe, fasteners, motors, machines, etc. In such cases reference is frequently made to one of the Renard number series. For further information on this point see Chapter 10, "Moving Into the World of SI."

Module. A module is an item with convenient dimensions or combination of dimensions, which by direct linear multiplication or subdivision gives other conveniently dimensioned items.

Modules most frequently occur where "add-ons" or "take-aways" are applicable to the total system, as in building construction.

For instance, the basic module for construction, recommended by ISO-R1006, is 100 mm (slightly less than 4 inches). For horizontal multi-modules ISO recommends 300 mm which is conveniently divisible by 2, 3, 4, 5, 6, 10, 12, 15, 20, 25, etc. The 300-mm multi-module is only slightly less than one foot (direct numerical conversion for one foot would be 304.8 or 305 mm).

1.13 Metric Module Demonstration and SI File

One of the most important first steps in preparing to work in SI is to develop a "new sense of dimension." This is best done by personally constructing a "metric module" chart. It is sometimes not enough to look at a printed chart, or at a chart constructed by others. Far more effective is to have the actual experience of making one's own chart. It can readily be done in a normal-size file folder which is generally 300 mm wide. The numerous opportunities for integer sub-modules may be easily illustrated as follows:

1. Cut out square pieces of contrasting colored paper sized 300, 150, 100, 75, 60, 50, 30, 25, 20, 15, 12, and 10 mm.
2. Paste these squares consecutively on the fully opened inside surface of a standard letter-sized file folder (approximately 11 3/4 inches wide), with one centerline of the 300-mm square coincident with the centerfold of the file folder and placing the upper-left-hand corner of each square coincident with the previous larger square in size sequence.
3. Draw horizontal lines across the folder from some of the lower edges such as the 12, 15, 25, 50, 60, 75, 100, and 150-mm squares and number the mm values at key points as illustrated in Fig. 1-5.

Ch. 1   INTRODUCTION TO SI   29

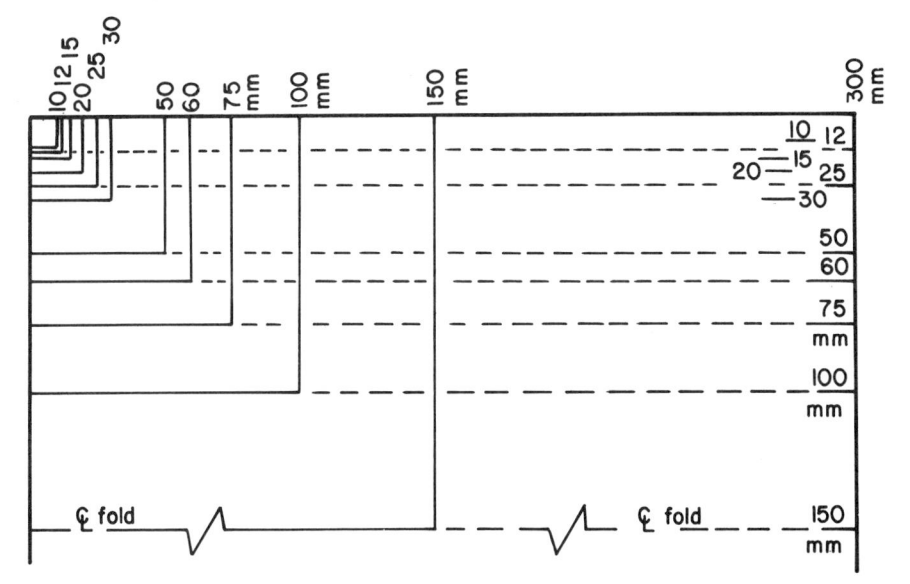

Fig. 1-5. Metric Module and Sub-Modules. (Partial illustration—not to scale.)

It is now possible to visualize exactly from this scaled Metric Module, combinations of various thicknesses, widths, and/or lengths; 300 × 10 mm; 60 × 25 mm; 50 × 15 mm; etc. Proportions illustrated in Fig. 1-5 can, of course, also be used to visualize larger units by direct decimal multiplication. Furthermore, the Metric Module folder in itself provides a ready and handy new reference file for metric (SI) memoranda.

NOTE—An alternate to some, or all of the "paste-up" plans is to draw the mm outline directly on the folder and then, if desired, apply further color with crayon or felt marker, especially in 10, 12, 15-; 20, 25, 30-; and 50, 60, 75-mm ranges.

2

Statics

2.0 Definition of Quantities

Before embarking upon the discussion of Statics it probably will be beneficial to reflect again and review in explicit language the definitions of the key engineering terms: mass, force, weight, load.

Unfortunately these terms are all-too-frequently used interchangeably in general conversation. In engineering work important distinctions must be made in accordance with the following technical definitions:

Mass of a body is the amount of matter which it contains compared to that of a standard prototype. Mass does *not* vary as a body is moved from place to place; it is a constant characteristic. Mass is properly defined as the resistance of a body to motion or to change in motion, in any direction. Mass is the inertial characteristic of a body. In SI terms the unit of mass is the kilogram.

Force is that physical phenomenon which, if unbalanced, will cause a mass to move or to change motion, or because of which motion impends. If the force is continuously applied, and remains unbalanced, the motion will change or will occur, and the body will be accelerated (or decelerated) according to Newton's Second Law:

$$F \propto ma$$

In SI terms the unit of force is the "newton" and is that force which will give to a mass of one kilogram an acceleration of one meter per second squared.

The phraseology "meter per second *squared*" is preferred since it parallels the symbolization of m/s².

Note that the motion described in $F \propto ma$ is independent of direction; i.e., it may be either horizontal or vertical, or at any other angle, except

Table 2-1. Quantities and SI Units Frequently Used in Statics

Quantity Symbol	Quantity Name	Unit Name	Unit Symbol	In Terms of Base Units
m	mass	kilogram	kg	kg
L, a, b	length, distance	meter	m	m
t	time	second	s	s
θ	angle	radian	rad*	...
F, P, Q, R	force	newton	N	kg·m/s²
M	moment	newton-meter**	N·m	kg·m²/s²
M_t	torsional moment	newton-meter**	N·m	kg·m²/s²
a	acceleration, linear	...	...	m/s²
g	acceleration, gravity	...	...	m/s²
W, w†	force, gravitational	newton	N	kg·m/s²

* Degrees (°), or slope ratios, are most frequently used in Statics.

** Force multiplied by perpendicular distance from center of rotation. Note that this is a simple static product (N·m) and can *not* be construed in terms of joules until movement occurs and work is done. In the latter case the principles of dynamics are applicable and displacement in the direction of the force (meters of linear displacement or radians of angular displacement) becomes involved.

† Some readers may prefer to avoid the use of this symbol because of the gravitational unit system inference. However, if W and/or w are used to denote "quantities" in SI it must be done with the understanding that the SI definition of this Quantity is "gravitational *force*" and that the units shall be newton (N).

that the motion and acceleration described are in respect to the line of action of the unbalanced force.

Note also that "force" is a vector quantity and as such must be described completely by: (1) a line of action; (2) point of application to the body; (3) magnitude; and (4) "sense" or direction along the line of action.

See Table 2-1 for a summary of "Quantities and SI Units Frequently Used in Statics."

In all cases of "Statics" the forces acting on a body are balanced and the body is at rest, but the characteristics are as described above.

Weight technically is a particular type of *force* which results from gravitational attraction. This force acts through a point customarily referred to as the c.g. Actually this point of action is the centroid of mass and should more properly be referred to as the "c.m." The line of action is always vertical and the direction is always downward. The magnitude of the force will depend on the local acceleration of gravity and on the amount of mass.

Since SI is a coherent system with no duplication or overlap of units there is no unique unit for weight. Technically the characteristics of weight are the same as those of any force. Thus the force of gravitational attraction on a body is always to be expressed in terms of "newtons."

In SI terminology a "weighing machine," "scale," or "balance" becomes a device for "ascertaining the mass" of a body by comparison with a standard mass. The result of such an operation, of course, is to be expressed in kilograms.

> *Load* is a *force* acting on a structure or machine and may be of various forms: live load, dead load, impact, wind, vibration, etc. All of these in SI should be expressed in "newtons" or in multiples thereof. As previously stated the load as measured by a testing machine, dynamometer, proving ring, test cell, or other force-determination device should be calibrated for read-out in newtons. A so-called "torque" wrench (a device that actually measures torsional moment) should have a numerical read-out based on newton-meters (N·m).

2.1 Absolute versus Gravitational System

The foregoing illustrates the fact that SI is an "absolute" system dealing basically with "mass" in all engineering considerations. That is, the potential acceleration due to gravity must be *inserted* into any determination where forces result from gravitational action.

This contrasts with the "customary" engineering system where gravity is normally presumed in all computations and only *taken out* where it is not applicable. Using such a procedure is described as working in a "gravitational" system.

The commonly used customary expression, $m = W/g$, is inconsistent with SI. To paraphrase a popular saying: "mass is mass, is mass!" And mass should always be expressed directly in kilograms. It is, of course, technically correct to state that $W = mg$, but *not* vice versa. Since there is a Base Unit for *mass* and a Derived Unit for *force* each should be viewed separately and the proper units of kilogram and of newton should be respectively applied.

The choice of a unit of force (newton) based on unit acceleration transforms a wide range of formulas by eliminating "g" where it formerly appeared and introducing it where formerly it was absent. The change is a sound and logical one but will require alertness and a new insight to force-mass relationships for those who have long been at home with a gravitational system.

Ch. 2   STATICS   33

The full significance of most of the above will be more apparent from the following review commentary with illustrations.

2.2 Understanding Force

The new unit of force is the starting point of our examination of the effect of SI units upon the subject of Statics. The concept of force may justly be described as the foundation stone of engineering as an exact science. It was probably first appreciated through the ever present necessity of resisting gravity; not until much later did it become associated with inertia and acceleration. While the subject of Statics deals only with forces in equilibrium, and makes very limited demands on units, it is the best starting point. Consider the elementary relationships demonstrated in Example 2-1.

Example 2-1

A simple windlass has a drum of radius r and an operating arm of length b. A horizontal force P is applied to the arm but does not succeed in moving it. What is the force F in the rope if the arm makes an angle θ with the vertical, and the rope is horizontal?

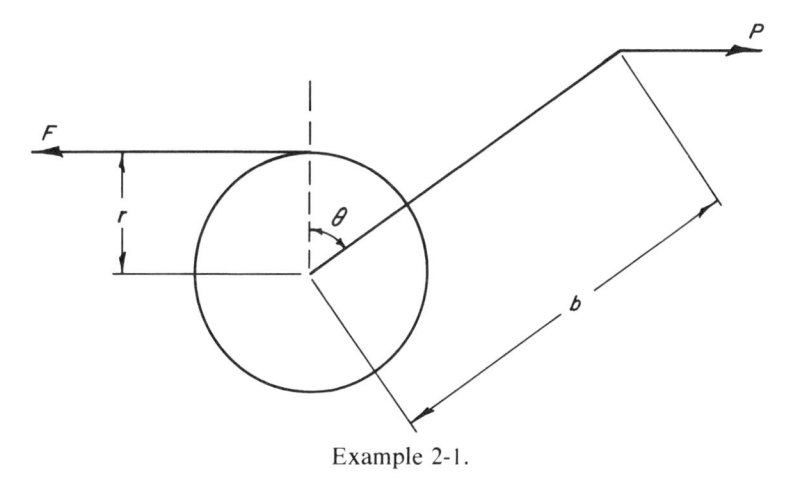

Example 2-1.

By taking moments about the shaft we find that:

$$P(b) \cos\theta = F(r), \quad \text{whence,} \quad F = \frac{P(b) \cos\theta}{r}$$

It will be seen that this expression utilizes three physical concepts—force, length, and angle; the need seldom arises in Statics for units of any other physical quantities.

2.3 Angular Measure in Statics

Since the direction of a force is one of its key characteristics, the matter of angular measure frequently arises in Statics. However, these angular relationships are given most often in terms of degrees or relative slopes. Thus, the natural trigonometric functions of angles are most serviceable in statics. The radian, preferred SI unit, is used chiefly in Dynamics.

2.4 Force of Gravity

The concept of *force* is inextricably bound up with the phenomenon of gravity and many calculations in Statics are concerned with supporting mass. Accordingly it seems advisable to examine in some depth the significant change in procedure which must be followed when SI units are used.

2.5 Newton's Laws

Newton's well-known First Law of Motion states that a body in motion (or at rest) continues in motion (or at rest) until acted upon by an external force. The Second Law states that the force required to accelerate (or decelerate) the body, or to change the direction of motion, is directly proportional to the mass. Engineers are also familiar with the three basic physical concepts linked by this Second Law of Motion: force, mass, and acceleration. As previously stated this law applies to *all* motion whether horizontal, vertical, or intermediate. The concept of mass has nothing directly to do with gravitation. Mass relates more closely to inertia which is multidirectional and independent of the intensity of the gravitational field. Astronauts have put it this way: "Walking on the moon is easier than on earth but stubbing your toe hurts just as much"!

Since the gravitational force exerted on, or by, a body is directly proportional to its mass and since the intensity of the gravitational field on the earth's surface varies by only about ½ percent, it has been customary to assign a standard value to "g" and thus to regard mass and weight as relatively interchangeable terms. Such liberty cannot be

Ch. 2 STATICS 35

taken generally. Nowadays every schoolboy knows that even a body under so-called "weightless" conditions still has the same mass as when directly subjected to gravity alone.

2.6 The Great Ambiguity

Until very recently such matters were the province only of the astronomer or the fiction writer; practical people, including engineers, were content to use terms such as "light" and "heavy" to describe mass and weight indiscriminately, to use a word such as "pound" to denote units of mass and of force. A mass of one pound, it was said, had a weight of one pound. The fact that two distinct concepts were involved was glossed over and, indeed, scarcely appreciated at all by many people. Going SI thus provides a timely opportunity for an essential clarification in engineering technical language.

The consequences of the present ambiguous customary practice may be seen by applying Newton's Second Law and taking care to discriminate between mass and force. For such discussion it is well to follow the convention of using "lb" to denote the pound-mass, and "lbf" for the pound-force. Assume that the law is represented by the equation:

$$\text{force} = \text{constant} \times \text{mass} \times \text{acceleration}$$

$$F = c \cdot m \cdot a$$

in which c is a proportionality factor.

Now consider the motion of a body of one lb when falling freely under the earth's gravity—that is, when subjected to the unresisted force of gravity on its mass, which is the force of gravity on one lb mass, and is called lbf.

Such a body experiences an acceleration of about 32.2 ft/s² so that the equation becomes:

$$1.0 \text{ lbf} = c \, (1.0 \text{ lb}) \, 32.2 \text{ ft/s}^2$$

It follows that the constant of proportionality must be:

$$c = \frac{1.0 \text{ lbf}}{(1.0 \text{ lb}) \, 32.2 \text{ ft/s}^2}$$

Disregarding the units in this expression, the equation in the general form is:

$$\text{Force} = \frac{1}{32.2} \times \text{mass} \times \text{acceleration}$$

2.7 The Engineer's "g"

Since the symbol "g" is commonly used to denote the gravitational acceleration, this expression is usually written:

$$\text{Force} = \frac{1}{g} \times \text{mass} \times \text{acceleration}$$

This is the form familiar to most engineers. It assumes that mass is expressed in terms of the lb, and force in terms of the lbf. This equation, in fact, constitutes a definition of the lbf unit, in which g must accordingly be given an agreed, exact value rather than the slightly variable value it has over the earth's surface.

In a precisely similar manner the metric engineer has been accustomed to define the kilopond (sometimes called a kilogram-force) using metric units of mass and of acceleration.

It is important to note that in both cases, the units of "g" are *not* ft/s² or m/s²; these g values are not, in fact, the acceleration due to gravity, though numerically they are the same as the *standard* value; their true units are

$$\frac{\text{lb} \cdot \text{ft/s}^2}{\text{lbf}} \quad \text{or} \quad \frac{\text{kg} \cdot \text{m/s}^2}{\text{kgf}},$$

respectively, and to make this quite clear these constants are sometimes denoted by a special symbol: g_c. The numerical values are, respectively, 32.1740 in U.S. Customary units and 9.80665 in metric units.

2.8 The Direct Approach to Force in SI

The approach to force units in SI is directed towards ensuring that the proportionality constant in Newton's Second Law is equal to unity. Thus:

$$\textit{Force} \text{ (SI unit, N)} = 1.0 \times \textit{mass} \text{ (kg)} \times \textit{acceleration} \text{ (m/s}^2\text{)},$$

then clearly, for consistency, the SI unit for force must be the kg·m/s². For greater convenience it is called the newton (N), but the full identification in terms of Base Units is equally correct.

Suppose this new equation is applied to the mass of 5.00 kg supported by a string, where the local value of the gravitational acceleration is 9.81 m/s²; then the equation becomes:

$$\textit{Force} = 5.00(9.81) = 49.05 \text{ N} \qquad \text{kg} \cdot \text{m/s}^2 = \text{N}$$

Ch. 2　　　　　　　　　STATICS　　　　　　　　　37

thus, the force of gravity on a mass of 5.00 kg at this location is 49.05 newtons, exactly.

2.9 Gravitational Force Determination

The general rule for determining the force of gravity, in SI units, on a given mass may be stated thus: take the mass in kg, and multiply it by the local acceleration of free fall expressed in m/s^2. The answer is the force effect of gravity in newtons. For almost all practical purposes on the earth's surface the standard value may be used, namely, 9.80665 m/s^2; usually 9.81 or even 9.8 is precise enough. If a 2 percent error is acceptable, one may even use 10 and save much arithmetic. The value of 9.8 will be used generally throughout this book for illustrative purposes only. Each user must decide on the correct value to be applied in accordance with the necessary precision of his work.

2.10 SI Procedure

The operation of the SI procedure will be explained by means of worked examples. First, however, the sizes of the relevant units will be compared with those of the familiar English units. The English system includes a variety of units for force, interrelated by many non-decimal numbers (1.0 long ton = 2240 pounds, etc.). In SI there is only one primary unit of force, namely, the newton (N). But unit multiples, interrelated by powers of 10, are available by use of the metric prefixes. This is, of course, one of the standard features of SI.

The range of magnitudes of these unit multiples brings into focus the point made earlier about the changed numerical values. Thus the newton (N) equals approximately 0.2248 lbf, and therefore may be accepted as a convenient size wherever the lbf has hitherto been used. That is to say, it is of a similar order of magnitude. But familiar numbers will obviously change significantly.

For larger forces the tonf is often used, but since the newton equals 0.0001124 tonf, a prefix should be applied to give a unit of convenient size. The kilonewton (kN) which equals about 0.1 tonf, and the meganewton (MN), about 100 tonf, will find common use in SI. It may be noted that in this instance the numbers appearing in formulas will be almost unchanged except for the location of the decimal point; although this is fortunate, it is just by chance.

The ounce-force is occasionally used in light engineering; since 1.0 N = 3.597 ozf, the newton will probably be found convenient as it stands, but one can also use the millinewton (mN) which equals 0.003597 ozf for extremely small forces. The full series of metric force units with prefixes will cover a far wider range than is likely to be needed—from 2.2×10^{17} to 2.2×10^{-19} lb, i.e., from 10^{18} N to 10^{-18} N, (EN to aN).

2.11 Comparative Example

Now consider again the example of the windlass presented previously, and suppose, for simplicity, that $\theta = 0°$ so that $\cos \theta = 1.0$. Suppose further, that there is a force $P = 50$ lbf, and it is determined by measurement that $r = 8$ in. and $b = 3$ ft 9 in. To perform the calculation necessary for finding F the first step would be to convert b to 45 in., whereupon $F = (50)\, 45/8 = 281$ lbf.

The same procedure conducted in SI units would appear as follows (the numbers are intended to be similar but *not* exact equivalents): applied force $P = 220$ N, $r = 200$ mm, $b = 1130$ mm.
Or, alternatively: $F = (220)\, 1.130/0.200 = 1243$ N.

2.12 SI Simplifies

Two points emerge from this comparison. First, the simple procedure of changing from a dimension quoted in terms of two different units, the foot and the inch, into one based on a single unit involves the factor 12 and a little mental arithmetic. The process is much more complicated when fractions of an inch are involved. On the other hand the corresponding process in metric units does not introduce any arithmetic since the factor is 1000. It is one of the cardinal features of the metric system that it does away with a great deal of the petty arithmetic which is inseparable from the English system of units. This may be regarded as a trivial observation but it is well to make it occasionally.

The engineer does not normally regard this preliminary simplification of units as forming part of the real calculation. When dealing with length, he works throughout in inches. Either by measuring directly in inches, or by instant mental arithmetic, he would pose the above problem in the form $P = 50$ lbf, $r = 8$ inches, $b = 45$ inches, whence $F = (50)45/8 = 281$ lbf.

The corresponding SI form would be $P = 220$ N, $r = 200$ mm, $b = 1130$ mm, (since the preference generally is for the millimeter).

This reveals the second point, namely, that these two calculations are identical in form; only the units and numbers are changed. This chapter on Statics might almost end at this point with the observation that, provided the proper definition of each quantity is observed and provided

Ch. 2 STATICS 39

that calculations are carried out consistently in terms of a single set of units, principles utilized in English units are equally applicable to SI units. It will nevertheless be useful to establish confidence by means of a few more worked examples.

Example 2-2

Find the resultant of the co-planar forces shown in the diagram.

Apart from the fact that the forces are expressed in newtons, there is nothing whatever novel about this calculation. By the usual cosine formula, the resultant is:

$$R = \sqrt{(22)^2 + (17)^2 + 2(22)(17)\cos 30°} = 37.7 \text{ N}$$

and

$$\tan \theta = \frac{22 \sin 30°}{(22 \cos 30°) + 17} = 0.305 \qquad \theta = 17.0°$$

Example 2-2

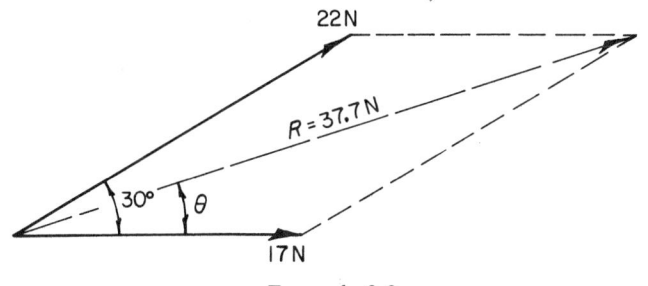

Example 2-3

Find the resultant of the concurrent, co-planar forces shown in the diagram.

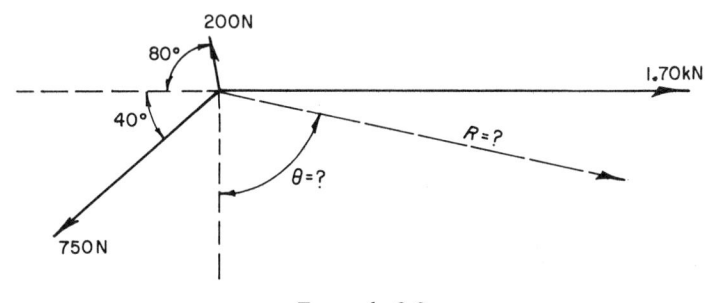

Example 2-3

Observe the "mixed" units: two of the forces are in newtons and one in kilonewtons. The first step should be to produce uniformity either by treating the forces as 750 N, 200 N, and 1700 N; or as 0.75 kN, 0.2 kN, and 1.7 kN. There is no arithmetic involved in this step in either case, and it would usually not even be written down; the choice depends entirely on an assessment of the relative convenience of the numbers involved. If it is decided to work in terms of the kilonewton, the resolution of all forces in the direction of the 1.7 kN force is:

$$1.7 - 0.75 \cos 40° - 0.2 \cos 80° = 1.09 \text{ kN}$$

and at right angles,

$$0.2 \sin 80° - 0.75 \sin 40° = -0.285 \text{ kN}$$

whence

$$R = \sqrt{(1.091)^2 + (-0.2851)^2} = 1.1276 \text{ kN, or } 1128 \text{ N}$$

and

$$\tan \theta = 1.091/0.2851 = 3.83; \quad \text{and } \theta = 75.4°$$

The sole element of novelty in this calculation lies in the manipulation of a prefix.

Example 2-4

Example 2-4.

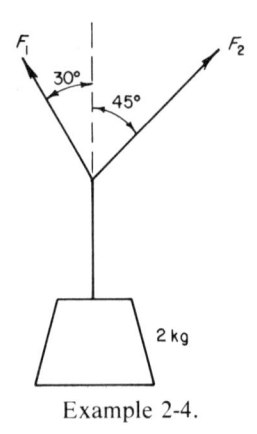

A body weighing 2 kg is supported by two strings as shown. Find the tensions in the strings.

Do not be put off by the loose phraseology in the question. The body does not "weigh" 2 kg in the strict sense; the unit stated shows unquestionably that it is the mass of the body which is given. To find its effect, which is required for the solution of the problem, it is necessary to multiply the mass by the local gravitational acceleration. Assume that 9.8 m/s² is accurate enough, then the force of gravity is:

$$2(9.8) = 19.6 \text{ N} \qquad \text{kg·m/s}^2 = \text{N}$$

Resolving vertically,

$$F_1 \cos 30° + F_2 \cos 45° = 19.6;$$

and horizontally,

$$F_1 \sin 30° - F_2 \sin 45° = 0,$$

whence

$$F_1 = 14.35 \text{ N} \quad \text{and} \quad F_2 = 10.15 \text{ N}.$$

These tensions, results of gravity, are in newtons. It would be quite incorrect to try and convert them into kilograms. A tension is a force and must be stated (and measured) in newtons.

2.13 Moment

The concepts of both bending moment and torsional moment involve the product of a force and a perpendicular length. The primary SI unit is accordingly the newton-meter. Various methods would seem acceptable for expressing such a "product" unit in abbreviated form as: N·m, N-m, N m, and Nm. As a matter of principle the product may equally well be written m·N, m-N, and mN. But in SI particular care is necessary when a unit symbol is similar to a prefix symbol. In this instance, "m" may refer either to the unit of "meter" or to the prefix "milli." There is no means of telling, apart from the context, whether mN means millinewton or meter-newton. It is therefore desirable to choose the order to avoid any possibility of confusion, and the meter-newton forms (m-N, etc.) should *not* be used for moment. It may then be assumed that mN always means "millinewton."

Since the Base Unit form of the newton (N) is the kg·m/s², the newton-meter (N·m) may, of course, be expressed also as: kg·m²/s².

2.14 Compound Units—Single Prefixes

Both newton and meter may be provided with prefixes. There is nothing wrong, for example, with N·mm or kN·m. Nor is there anything wrong with kN·mm either, but this is only a needlessly long way of writing N·m, since the kilo and the milli cancel out. It is useful to appreciate that a compound unit like this is a unit in its own right. If a moment is produced from a force, measured in kilonewtons; and a length, measured in millimeters, the resulting unit should *not* be thought of merely as the product of kN and mm, with the two units regarded as existing quite separately. The primary unit for moment is the N·m; nothing is achieved by arguing about where to put any prefixes that may appear appropriate.

A prefix is merely a shorthand way of writing down, in as direct a manner as possible, the power of ten which should be applied to the root product; hence, N·m, kN·m, MN·m, etc., are indeed the preferred forms.

Example 2-5

Express the system of parallel forces shown as a single force through the point 0, together with a couple.

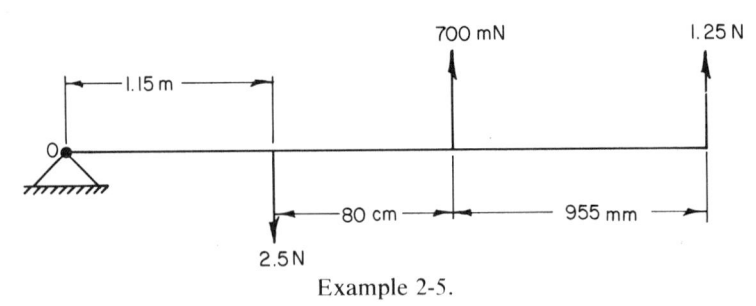

Example 2-5.

The prefixes here are mixed in a most untidy way, and this is *not* intended as an example of how to present such a problem. Nevertheless it offers no difficulty as the necessary adjustments can be made on inspection.

The resultant force is $(2.5 - 0.7 - 1.25)$N $= 0.55$ N

The total moment about 0 is:

$$M_0 = 2.5(1.15) - 0.7(1.15 + 0.8)$$
$$- 1.25(1.15 + 0.8 + 0.955) \text{ N·m}$$
$$M_0 = 2.5(1.15) - 0.7(1.95) - 1.25(2.905)$$
$$= -2.121 \text{ N·m}$$

These calculations have been carried out in terms of the primary units:

Ch. 2　　　　　　　　　　　STATICS　　　　　　　　　　　43

meter, newton, and newton-meter. This is not essential, of course; what matters is that the selected units should be used consistently. For example, a common practice is to use the millimeter rather than the meter, as the millimeter is the unit commonly used for engineering dimensions of machinery.

Example 2-6

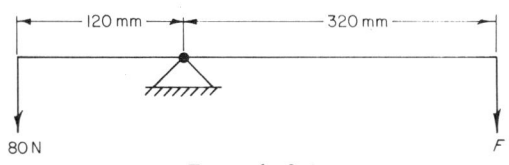

Example 2-6.

What force F is required to balance the lever shown?
Taking moments about the fulcrum:

$$80(120) = F(320) \quad \text{N·mm} = \text{N·mm}$$

so that

$$F = 30 \text{ N}$$

Example 2-7

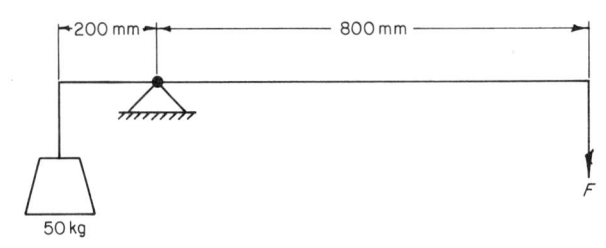

Example 2-7.

What force F is required to provide equilibrium for the 50 kg "mass"?
Once again, when in doubt about weight, mass, and force examine the unit: kg means mass. The force of gravity is:

$$50(9.8) = 490 \text{ N};$$

whence:

$$F = \frac{(490)200}{800} = 122.5 \text{ N}$$

Example 2-8

Suppose the previous example was a diagrammatic representation of a balance; what "weights" (or "weight-pieces") would be needed on a scale-pan to produce F?

As already shown, the force at F must be 122.5 N; evidently this could be produced, in the form of weight, by putting on the scale-pan a mass of $122.5/9.8 = 12.5$ kg. This might well be labeled a "weight" in common terms. This calculation involved first multiplying by 9.8 to get weight and then dividing by 9.8 to get mass. It is obvious that this is unnecessary; all that is needed is to write:

$$\text{mass required in scale pan} = 50(200/800) = 12.5 \text{ kg}$$

The short-cut nature of this calculation should be appreciated. Thus, there is no real significance to any implied unit of "kg·mm" which apparently has been used. This calculation is a simple matter of ratios.

2.15 Ascertaining Mass

There is much undiscerning argument about "weighing" and the use of "weights" on scale-pans. It is sometimes said that, since in using a "weight" one is employing its gravitational force rather than its mass, the artifact should be calibrated and labeled in newtons. This betrays a lack of appreciation of the true nature of the situation. The mass of a body is the same wherever it may be; the weight of a body, i.e., the force exerted upon it by the local gravitational field, varies with its position. It would be in order to label a body in terms of newtons only if it could be guaranteed that the body would never be moved to a new position where the gravitational field was different and only if its use and application were to be restricted always to *gravitational* situations.

2.16 Mass-Force Examples

In exactly the same way it is incorrect in the U.S. Customary system to label an artifact in terms of lbf.

Hence, any artifact or body should be labeled in terms of its mass which is constant, i.e., in terms of lb or in terms of kg.

In setting out to "weigh" a body the ultimate purpose is to know its mass or its weight, or indeed both. As in most measurements, the procedure is to compare the body with a standard body or bodies; this is the process idealized in Examples 2-7 and 2-8; that is, weight in its true sense is used for the comparison, because it is so very much easier than

Ch. 2				STATICS				45

using mass. The validity of this comparison stems from the fact that both the body to be weighed and the standard bodies are in the same gravitational field. The procedure is therefore quite rightly "weighing" the body, but it is proper to *deduce* information about its mass from the operation. It is merely a colloquialism to say "the body was found to weigh 5 kg," when strictly speaking, it should be said that: "the body has a *mass* of 5 kg."

Example 2-9

Example 2-9.

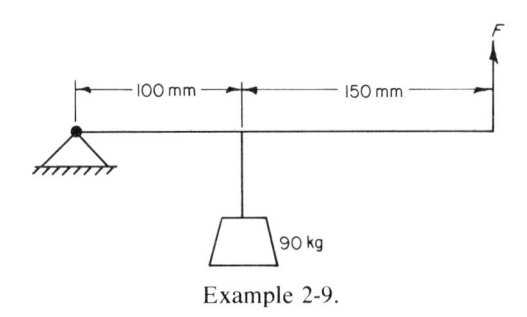

What force F is required to support the weight of 90 kg?
The force of gravity on the weight (mass) is:

$$90(9.8) = 882 \text{ N};$$

whence:

$$F = 882(100/250) = 352.8 \text{ N}$$

Note that in this instance there is no possibility of regarding F, itself, as being associated with gravitational attraction so that here the procedure indicated in Example 2-8 is not applicable.

Example 2-10

What force F is required to support the 2 t load?
The symbol t represents "metric ton," i.e., 1000 kg (the correct SI name is the megagram, or Mg).* This is a mass unit and the corresponding force is accordingly, $2000(9.8) = 19.6$ kN. Obviously then, $F = \frac{1}{2}(19.6)$ kN $= 9.8$ kN.

---

* It is interesting to note that for many years engineers have used the special term "kip" to represent the convenient quantity of 1,000 lbf in the U.S. Customary Unit System. SI automatically provides similar convenience in *regular* units with the kN, MN, etc.

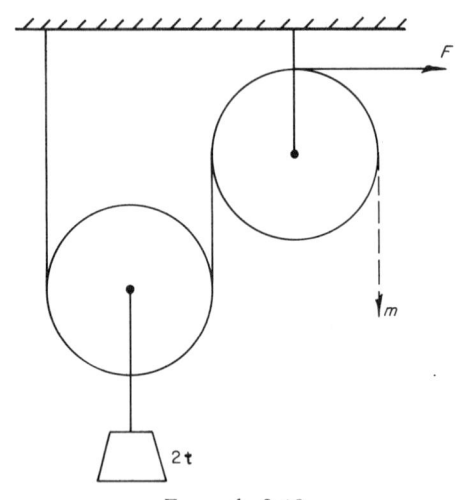

STATICS Ch. 2

Example 2-10.

If the force F were to be produced by a suspended mass, m, the shortcut, ratio procedure would give at once: $m = \frac{1}{2}(2.0 \text{ t}) = 1.0$ metric ton. When it is required to know the tension in the rope, however, the proper calculation is as indicated initially, since the tension must be given in newtons.

Example 2-11

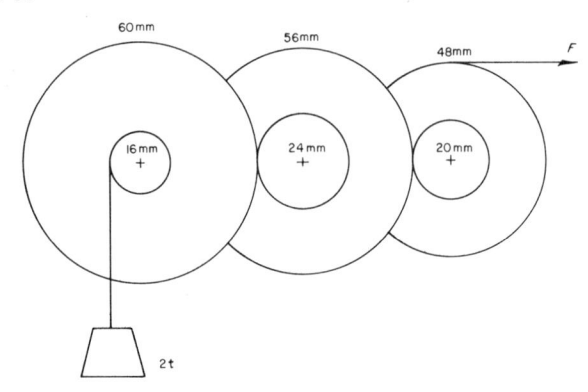

Example 2-11.

The pitch diameters of a set of gears and winding drums are as shown in the diagram. What force F will be needed to raise the mass of 2 metric tons?

Ch. 2   STATICS   47

The "weight" (mass) of 2 t represents a force of 2000(9.8) = 19 600 N. Accordingly:

$$F = 19\,600 \times \frac{16}{60} \times \frac{24}{56} \times \frac{20}{48} = 933 \text{ N}$$

This force could, of course, be produced alternately by suspending a mass of 933/9.8 = 95 kg from a wraparound rope. This latter solution suggests that the required mass for such suspension could be obtained by direct multiplication as in the customary gravitational system, but it should be noted that this procedure does not give the answer in SI force units, as is desired for a more general analysis.

Example 2-12

A screw jack has a square thread having a 5-mm pitch and a 50-mm pitch diameter. If a force of 80 N is applied at a radius of 750 mm on the handle, and there is no friction on the thread, what force must the threads support and what weight (mass) will the screw raise?

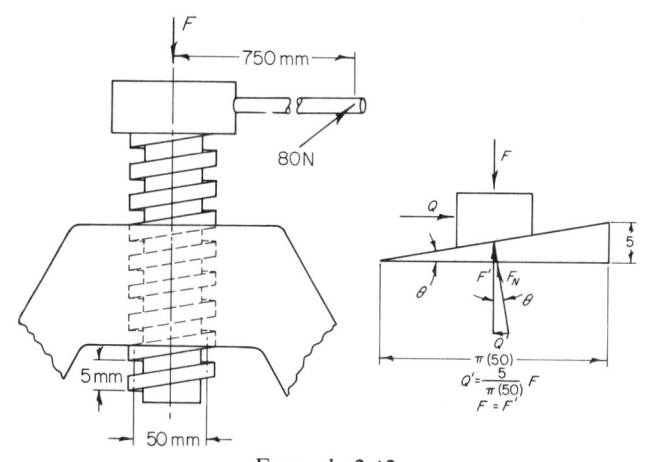

Example 2-12.

The force, Q, acting at the pitch diameter of the thread, causes the screw to move upward. This force is determined by the principle of moments about the axis of the thread.

$$\frac{50}{2} Q = 750(80)$$

$$Q = 2400 \text{ N}$$

48   STATICS   Ch. 2

The relationship of the forces may be viewed as similar to those which accompany equilibrium on an inclined plane, where the effective slope is equal to the ratio of the pitch and the pitch diameter. An equilibrium force, Q', resulting from the force due to the mass, is equal and opposite in direction to the force, Q.

$$Q' = \frac{5}{\pi(50)} F$$

Thus:

$$2400 = \frac{5}{\pi(50)} F$$

From which:

$$F = 75\,400 \text{ N, or, } 75.4 \text{kN}$$

Therefore, the threads are supporting a force of 75.4 kN. Strictly speaking, the latter part of the question has already been answered, namely: a weight causing a force of 75.4 kN is raised by the jack. However, interpreting the question to mean "What mass will the screw raise?" leads to the answer:

$$\frac{75\,400}{9.8} = 7690 \text{ kg} = 7.69 \text{ Mg, or, } 7.69 \text{ t}$$

Example 2-13

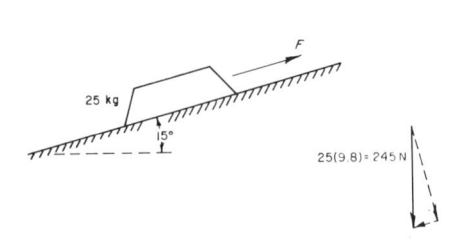

Example 2-13.

A body having a mass of 25 kg rests on a 15° slope. What force is required to begin to move it up the slope if friction is neglected?

Resolving forces parallel to the slope, and noting that the gravitational (vertical) force effect of the body is:

$$P = 25(9.8) \text{ newtons}$$

and the result obtained is:

$$F = 25(9.8) \sin 15° = 63.4 \text{ N}$$

Example 2-14

Re-calculate Example 2-13, assuming that the coefficient of friction between the body and the plane is 0.1.

The procedure for determining the frictional force between two bodies is to multiply the normal force between them by the appropriate coefficient of friction. The normal force is measured in newtons and the coefficient is merely a ratio, therefore the product is in newtons. Resolving perpendicular to the plane gives the normal force:

$$P_n = 25(9.8) \cos 15° = 236 \text{ N}$$

The friction force is, therefore, 236(0.1) = 23.6 N. Resolving parallel to the plane gives the force required to move the body:

$$F = 63.4 + 23.6 = 87 \text{ N}$$

Example 2-15

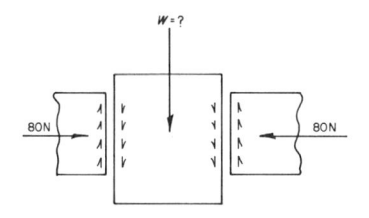

Example 2-15.

A block of metal is to be held between the jaws of a clamp, as shown, the faces being vertical; the force exerted by the clamp is 80 N. What is the greatest permissible "weight" (mass) of the block if the coefficient of friction is 0.7?

The block has two sides, so that the maximum frictional force is 2(0.7)80 = 112 N; this is the gravity force that can be resisted. It would correspond to a *mass* of 112/9.8 = 11.43 kg, which might colloquially be called the "weight" of the block.

2.17 Pressure

The "normal force" referred to in the above examples is sometimes loosely called the "pressure." This is an incorrect use of the term and is best avoided even in casual language, as it is apt to cause confusion and error. Pressure is force per unit area; in friction calculations it is the actual total force normal to the frictional surface that matters. Occasionally, the total force has to be calculated from the intensity of pressure multiplied by the area on which it acts.

The SI unit of pressure is derived from unit force divided by unit area, and is, accordingly, the N/m². Its correct use is illustrated in the following example.

Example 2-16

A hydraulic ram operates by oil pressure on a piston having a cross-sectional area of 2500 mm². What force will be exerted by the ram when the oil pressure is 2 MN/m²?

The required answer is the product of the pressure and the area on which it acts; i.e., $F = 2$ MN/m² $\times$ (2500 mm²). Replacing the prefixes by the powers of 10 they represent, and providing a separate unit checkout, we have:

$$F = (2 \times 10^6)(2500 \times 10^{-6}) = 5000 \text{ N}$$

$$\frac{\text{N}}{\text{m}^2} \cdot \text{m}^2 = \text{N}$$

Several points deserve mention here. The first is the appearance of a compound unit formed by dividing one unit by another. The newton per square meter may be written as N/m², $\frac{\text{N}}{\text{m}^2}$, or N·m⁻².

The second observation concerns the addition of a prefix to such a unit, to give the meganewton per square meter. The third point is the significance of mm². It is important to note that the exponent applies both to the unit and to its prefix, so that mm² = $(10^{-3}$ m$)^2 = 10^{-6}$ m².

2.18 The Pascal and the Bar

The unit for pressure, the N/m², may appear in several forms. As a widely used Derived Unit it now (since 1971) has been given the name "pascal," for which the symbol is Pa. Since the Base Units of the newton are kg·m/s², the Base Units of the pascal are: kg/m·s². The pascal is the preferred unit for unit pressure, for unit stress, and for modulus of elasticity.

Another special name, "bar," is sometimes used for 10^5 N/m²; this is *not* the preferred multiple in SI but it has had significant use, for example, in meteorology; in thermodynamics; and in fluid power where it is found both as the "bar" and as the "millibar." But this is outside of SI. Further use of the bar, or any other multiple, is not recommended, especially since the millibar introduces another anomaly, 10^2 Pa.

In some other instances it has been deemed desirable to use a unit having the millimeter in the denominator, i.e., the N/mm^2; this is, of course, exactly the same as the MN/m^2. The presumed convenience of this form arises largely from the practice of giving dimensions in terms of millimeters; for instance, in this example just given we could have written:

$$F = 2(2500) = 5000 \text{ N} \qquad \frac{\text{N}}{\text{mm}^2} \cdot \text{mm}^2 = \text{N}$$

cancelling out the mm^2 directly. The advantage is limited to particular situations and must be balanced against the overall advantages of consistency in terms of primary units.

As an exercise in SI terminology it may be noted that the MN/m^2 = MPa = N/mm^2 = 10 bar. The pascal is a very small unit; 1 Pa = 0.000 145 psi. For this reason units such as the kilopascal (1 kPa = 0.145 psi) and megapascal (1 MPa = 145 psi) are much more frequently encountered. From this it is evident that the principal attraction of the "bar" is that it is approximately equal to one atmosphere.

Having again reviewed all of the above, and other considerations, the 14th CGPM in 1971 adopted the special name "pascal" (symbol Pa) for the SI unit of pressure (newton per square meter). Through the further work of CIPM and ISO Technical Committee 12, the preferred multiples and sub-multiples have been identified as:

$$\text{GPa} \quad \text{MPa} \quad \text{kPa} \quad \text{Pa} \quad \text{mPa} \quad \mu\text{Pa}$$

2.19 Additional Examples

Following are a number of other examples, one of which is worked graphically to additionally illustrate the applications in SI of the principles of Statics. Other uses of these unit system principles and preferred SI units will, of course, be found in all of the succeeding chapters.

Example 2-17

A framework of rigid bars carries the loads indicated. What are the reactions at A and D?

Load at point F:

= 200(9.8) = 1960 N = 1.96 kN   kg·m/s² = N

Taking moments about point A:

$-1.96(6) - 8.00(2) - H_D(10) = 0$ $H_D = -2.78$ kN

By summation of forces, horizontally:

$-8.00 - 2.78 + H_A = 0$ $H_A = 10.78$ kN

By summation of forces, vertically:

$-1.96 + V_A = 0$ $V_A = 1.96$ kN

Example 2-17.

Example 2-18

A cylinder 4.00 m in diameter, having a mass of 30.0 kg is supported by a pin-ended rigid bar, AB, and a vertical cable at B, as shown in the accompanying sketch. Determine the required tension in the cable and the thrust against the wall at C.

The principles of static equilibrium may be applied in the manner indicated, yielding the answers in force units (newtons) as shown.

By geometry:

$$P_c = \frac{30.0(9.8)1}{\sqrt{3}} = 169.7 \text{ N}$$

$$P_D = 2P_c$$

$$P_D = 339.5 \text{ N}$$

$$\text{AD} = \frac{(2+1)2}{\sqrt{3}} = 3.464 \text{ m}$$

Ch. 2   STATICS   53

Taking moments about A:

$$3.464P_D = 6T_B$$

$$T_B = \frac{339.5(3.464)}{6} = 196.0 \text{ N}$$

Also by

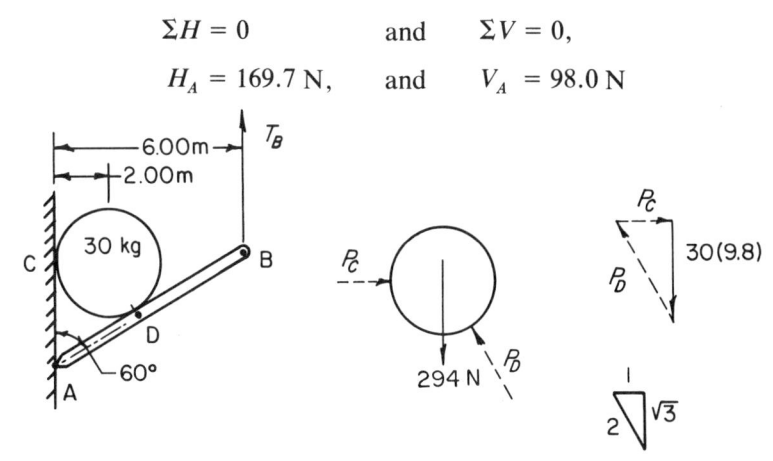

$\Sigma H = 0$ and $\Sigma V = 0$,

$H_A = 169.7$ N,   and $V_A = 98.0$ N

Example 2-18.

Example 2-19

Given a girder with the rigid arms shown and the applied loads as indicated, determine the required reaction components at the supports.

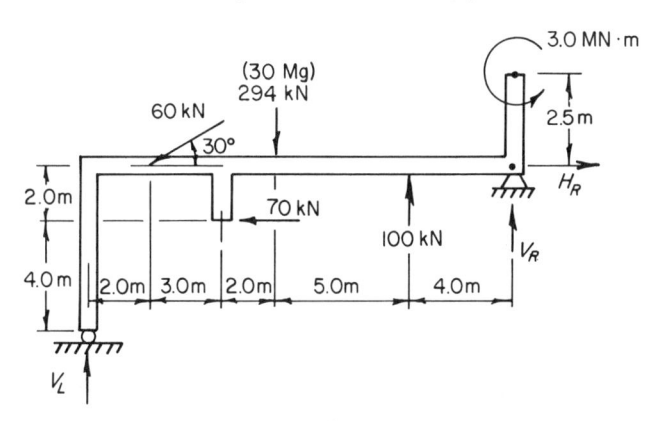

Example 2-19.

Evaluating force resulting from mass loading:

30 Mg = 30 000 kg   30 000(9.8) = 294 000 N = 294 kN

Summation of moments (in kN·m) about the right pin reaction equal to zero:

$$V_L(16) - 60 \sin 30°(14) + 70(2)$$
$$- 294(9) + 100(4) - 3000 = 0$$
$$16V_L - 420 + 140 - 2646 + 400 - 3000 = 0$$
$$16V_L = 5526 \text{ kN·m} \quad V_L = 345 \text{ kN}$$
$$\Sigma V = 0 \quad 345 - 60 \sin 30° - 294 + 100 + V_R = 0$$
$$V_R = 121 \text{ kN}$$
$$\Sigma H = 0 \quad -60 \cos 30° - 70 + H_R = 0$$
$$H_R = 122 \text{ kN}$$

Example 2-20

Given the pin-connected framework supporting the loads indicated in the accompanying sketch, determine the stresses in all members.

Evaluating mass loading:

3.0 Mg = 3.0 × 10³ kg   3.0 × 10³(9.8) = 29.4 kN

Example 2-20.

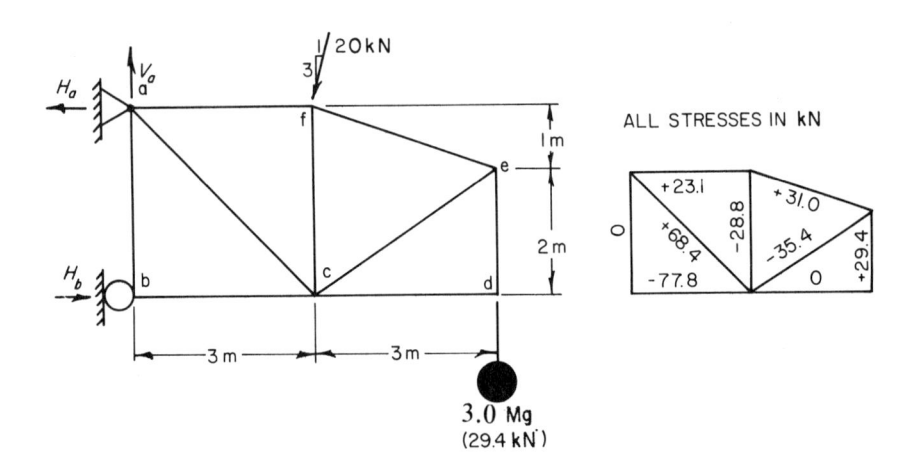

Ch. 2　　　　　　　　　　STATICS　　　　　　　　　　55

By the usual principles of statics the horizontal and vertical components of the inclined 20 kN force are 6.3 kN and 19 kN, respectively.
Solving for the reactions by the usual principles of statics:

$\Sigma V = 0$　　$V_a - 19 - 29.4 = 0$　　　　　　$V_a = 48.4$ kN

$\Sigma M_a = 0$　　$19(3) + 29.4(6) - 3H_b = 0$　　$H_b = 77.8$ kN

$\Sigma H = 0$　　$77.8 - 6.3 + H_a = 0$　　　　　$H_a = -71.2$ kN

With these reactions known the stresses can readily be determined by either algebraic or graphical means. The results, in kN, are tabulated.

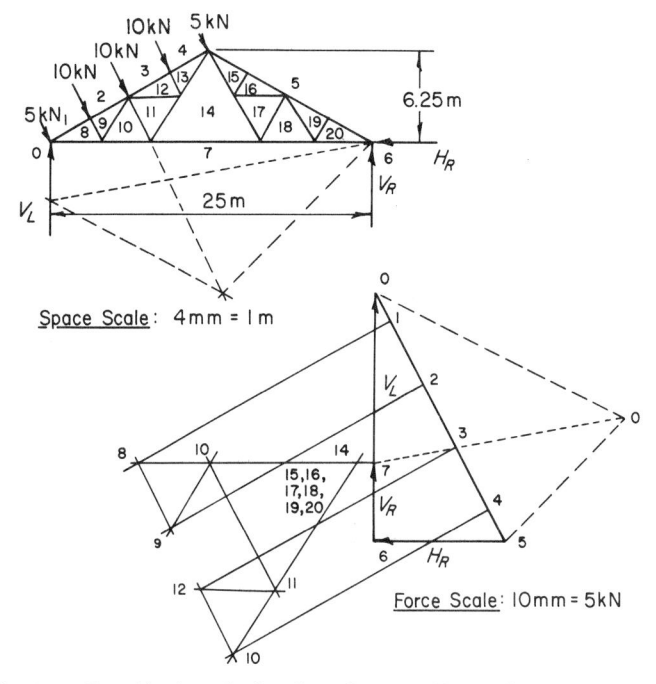

Fig. 2-1. Graphical analysis of roof truss (Illustrative; not to scale).

Example 2-21

Analyze the roof truss shown, to determine the stresses in all members for the wind load condition indicated in Fig. 2-1.

A complete graphical solution for reactions and stresses is also shown in Fig. 2-1. There is nothing unusual about this solution except that all forces are in kN and the graphical scales are naturally decimal, appropriately fitting the metric concept.

Reactions: Stresses:

$H_R = 19$ kN	1-8 = -41 kN	8-7 = +34 kN	8-9 = -10 kN
	2-9 = -41 kN	10-7 = +23 kN	12-13 = -10 kN
$V_R = 11$ kN	3-12 = -41 kN	14-7 = + 2 kN,	11-10 = -20 kN
	4-13 = -41 kN	etc.	10-9 = +10 kN
$V_L = 24$ kN			12-11 = +10 kN,
			etc.

Example 2-21.

2.20 Statics Revisited

In a sense this chapter may have appeared to carry to an extreme the discussion of the concepts of mass and force as applied to Statics. But with the marked change in measurement system approach that is occurring, it seemed advisable to review many facets. Perhaps it is opportune now, before moving on, to summarize as follows:

1. SI provides two discrete units which make a new, clear distinction between mass and force.
2. Whatever terminology is used in the statement of a given condition, the SI unit (kilogram for mass or newton for force) should be the determining factor in identifying the physical quantity described.
3. To avoid any further ambiguity it seems advisable at this time of changeover of measurement system, to de-emphasize or eliminate the term "weight."
4. More obvious benefits from the use of the newton as the force unit in a new, coherent measurement system will become increasingly apparent as the related units of pressure, energy, and power (pascal, joule, and watt) are further discussed in the following chapters involving Dynamics.

3
Dynamics

3.0 Motion

Dynamics is concerned with motion: the motion of individual particles and the motion of collections of particles bound together as in solids or fluids. Motion involves direction of movement along an identified path of action, as well as magnitude. The motion may be rectilinear, curvilinear, or rotational; it may include rolling and/or sliding.

Dynamic considerations include relationships between: mass, force, length, and time. Dynamics deals with: impulse, momentum, impact, vibration; work, energy, and power.

3.1 Velocity and Acceleration

The study of dynamics is based on two key concepts: (1) *velocity*, the time rate of change of position; and (2) *acceleration*, the time rate of change of velocity. Both of these concepts may be expressed either in linear or in rotational terms. Linear motion is generally referred to as translation. Rotation and translation must be treated separately; in SI units these two actions are expressed as follows:

Quantity	Translation Symbol	Translation Units	Rotation Symbol	Rotation Units
velocity	$\dot{V}$	m/s	ω	rad/s
acceleration	a	m/s^2	α	rad/s^2

Related information on "Quantities and SI Units Frequently Used in Dynamics" is given in Table 3-1. Formulas Frequently Used in Dynamics are given in Table 3-2.

Table 3-1. Quantities and SI Units Frequently Used in Dynamics

Quantity Symbol	Quantity Name	Unit Name	Unit Symbol	* In Terms of Base Units
m	mass	kilogram	kg	kg
t	time	second	s	s
L, s, δ	length, distance, displacement	meter	m	m
θ	angular displacement	radian	rad	...
$\bar{V}$	velocity, linear	...	...	m/s
ω	velocity, angular**	...	...	rad/s
a	acceleration, linear	...	...	m/s^2
α	acceleration, angular	...	...	rad/s^2
g	acceleration, gravity	...	...	m/s^2
F	force	newton	N	kg·m/s^2
T	torque	newton-meter per radian	N·m/rad	kg·m^2/(rad·s^2)
$\bar{I}$	inertia, rotational	...	...	kg·m^2/rad^2
Ft	impulse, linear	...	N·s	kg·m/s
$m\bar{V}$	momentum, linear	...	...	kg·m/s
Tt	angular impulse; (N·m/rad)s	...	(N·m)s/rad	kg·m^2/(rad·s)

$\bar{I}\omega$	angular momentum; (kg·m²/rad²) rad/s	...	kg·m²/(rad·s)
W	work; $(F\,s)$; (N·m)	joule	J	kg·m²/s²
	$(T\,\theta)$; (N·m/rad) rad	joule	J	kg·m²/s²
E_p	potential energy	joule	J	kg·m²/s²
E_k	kinetic energy	joule	J	kg·m²/s²
P	power; $(F\,s/t)$; (N·m)/s	watt (J/s)	W	kg·m²/s³
	$(T\,\theta/t)$; (N·m/rad) rad/s = (N·m)/s	watt	W	kg·m²/s³
	$(2\pi\,n\,T)$; $\dfrac{\text{rad}}{\text{r}} \cdot \dfrac{\text{r}}{\text{s}} \cdot \dfrac{\text{N·m}}{\text{rad}} = \dfrac{\text{N·m}}{\text{s}}$	watt	W	kg·m²/s³
r, d	radius, diameter	...		m
r	rotational factor (velocity)	...		m/rad
r	rotational factor (acceleration)	...		m/rad²
k	gyrational mass factor; $k = \sqrt{\bar{I}/m}$	...		m/rad
k	spring modulus		N/m	kg/s²

* For complete interpretation of this Table, note the following in terms of Base Units:

$$N = kg \cdot m/s^2 \qquad J = N \cdot m = kg \cdot m^2/s^2 \qquad W = J/s = kg \cdot m^2/s^3$$

** Usually observed or quoted in terms of "revolutions," i.e.: r/min; r/s.

Table 3-2. Formulas Frequently Used in Dynamics*

Number	Equation	Number	Equation
(3.1)	$F = ma$	(3.14)	$\omega = \omega_0 + \alpha t$
(3.2)	$\bar{V} = \bar{V}_0 + at$	(3.15)	$\theta = \omega_0 t + \frac{1}{2}\alpha t^2$
(3.3)	$s = \bar{V}_0 t + \frac{1}{2}at^2$	(3.16)	$2\alpha\theta = \omega^2 - \omega_0^2$
(3.4)	$2as = \bar{V}^2 - \bar{V}_0^2$	(3.17)	$\bar{V} = \underline{r}\omega$
(3.5)	$s = \frac{1}{2}(\bar{V} + \bar{V}_0)t$	(3.18)	$F_c = m\underline{r}\omega^2$
(3.6)	$Ft = m(\bar{V} - \bar{V}_0)$	(3.19)	$F_c = m\bar{V}^2/r$
(3.7)	$t = 2\pi\sqrt{m/k}$	(3.20)	$T = \bar{I}\alpha$
(3.8)	$t = 2\pi\sqrt{L/g}$	(3.21)	$\bar{I} = \frac{1}{2}m\underline{r}^2$
(3.9)	$W = Fs$	(3.22)	$\bar{I} = mk^2$
(3.10)	$E_p = mgh$	(3.23)	$Tt = \bar{I}(\omega - \omega_0)$
(3.11)	$E_p = \frac{1}{2}k\delta^2$	(3.24)	$W = T\theta$
(3.12)	$E_k = \frac{1}{2}m\bar{V}^2$	(3.25)	$E_k = \frac{1}{2}\bar{I}(\omega^2 - \omega_0^2)$
(3.13)	$P = Fs/t$	(3.26)	$P = T\theta/t = T\omega$

Note—For the purpose of simplification, all proportionality factors are customarily omitted from these formulas. Such factors depend upon the units used for the quantities involved. For a coherent system, such as SI, the magnitude of the proportionality factors is unity (1.0), but some factors may retain a unit characteristic. See especially, Section 3.18, "Circular Motion; Centrifugal Force" and sections that follow.

3.2 Mass and Gravity

As previously observed it is essential to keep in mind at all times that SI is an "absolute" system having a Base Unit of mass, the kilogram (kg). Thus, computations in SI treat the gravitational attraction in the same manner as any other force and require putting the value of g (usually 9.8 m/s²) *into* the calculation whenever gravitational effects are a factor influencing motion.

In this respect the use of SI contrasts sharply with the present "customary" procedures of engineering computations that generally assume a gravitational system. In other words, in SI the force of gravity is treated like any other force and the expression $m = W/g$ is *not* normally useful.

Another way of stating this distinction is to observe that in a gravitational system of units, such as the customary system, the base quantities are force, length, and time (usually written F, L, t) whereas in an *absolute* system, such as SI, the base quantities are mass, length, and time (m, L, t).

3.3 Force

As previously stated, force is that effect which tends to change the state of rest or motion of any body. When an unbalanced force acts on a body there will result an acceleration inversely proportional to the mass of the body and directly proportional to the magnitude of the unbalanced force. This statement gives the fundamental dynamic expression:

$$F \propto ma$$

If the units of force are properly selected the factor of proportionality becomes unity (1.0). For this reason, the SI unit of force, the *newton* (N), is defined as that force required to cause a unit mass of one kilogram (1.0 kg) to accelerate at the rate of one meter per second squared (1.0 m/s^2). Therefore:

(Eq. 3.1) $\qquad F = ma \qquad$ kg·m/s^2 = N

Such direct one-to-one relationships give SI the highly desirable characteristic of "coherence" and eliminate in later computations the many cumbersome unit interchange factors found in the "customary" system.

Since the newton is a rather small force, the magnitude of forces in engineering computations will be expressed most frequently in terms of the preferred unit: kilonewton (kN). Large forces (10^6 N, or over) will generally be expressed in meganewtons (MN), and very large forces (10^9 N, or over) in giganewtons (GN).

3.4 Unit Check-outs

In all engineering calculations it is always good practice to test the validity of numerical results by means of a check-out of units. This is especially true when utilizing a new unit system such as SI. Thus a major emphasis in these notes will be to include unit check-outs with most examples. In general, these check-outs are made in terms of Base Units but the selection is arbitrary and other sets are often convenient for use in checking.

3.5 Rectilinear Motion

In terms of the definitions of velocity and acceleration given earlier, the expressions given below, in various forms, may be used to solve problems of uniformly accelerated linear motion (translation):

(Eq. 3.2) $\bar{V} = \bar{V}_0 + at$

(Eq. 3.3) $s = \bar{V}_0 t + \frac{1}{2}at^2$

(Eq. 3.4) $2as = \bar{V}^2 - \bar{V}_0^2$

(Eq. 3.5) $s = \frac{1}{2}(\bar{V} + \bar{V}_0)t$

in which,

$\bar{V}_0$ = initial velocity   m/s

$\bar{V}$ = final velocity   m/s

a = linear acceleration   m/s^2

t = elapsed time   s

s = linear displacement   m

In some cases it will be necessary to resolve various movements into components taken in respect to given or conveniently selected axes, such as X-X, Y-Y, etc. In all cases it will be essential to check data and answers for consistency in units, as indicated in the following examples.

Example 3-1

What uniform acceleration is required to move a body on a smooth horizontal plane from a position of rest to a speed of 18 m/s in an elapsed time of 6 seconds? What will be the distance traversed?

Applying Equations (3.2) and (3.3), respectively:

$\bar{V} = \bar{V}_0 + at$ $\bar{V}_0 = 0$ *Unit Check*:

$a = (18 - 0)/6$ (m/s)/s = m/s^2

$a = 3$ m/s^2

$s = \frac{1}{2}(\bar{V} + \bar{V}_0)t$

$s = \frac{1}{2}(18 + 0)6 = 54$ m   (m/s)s = m

In Example 3-1, a unit check-out placed to the right in roman (upright) type has been carried through in parallel with the computations to demonstrate that the constant identification of units enables a check on the correctness of the application of the formulas. Normally, a simple example like this would not require units to be carried forward in such a punctilious manner, but in more complicated problems the unit check-out procedures are often of inestimable value as will be demonstrated in some of the examples which follow.

Ch. 3   DYNAMICS   63

Example 3-2

What velocity will result in applying a uniform acceleration of 4 m/s² for a period of 3 seconds to a mass already moving at 7 m/s in the same direction?

Applying Equation (3.2):

$$\bar{V} = \bar{V}_0 + at$$

$$\bar{V} = 7 + (4)3 = 19 \text{ m/s} \qquad (\text{m/s}) + (\text{m/s}^2)\text{s} = \text{m/s}$$

Example 3-3

The maximum speed of a train operating on a certain stretch of track is 72 km/h. How far back from a danger point should a warning signal be placed if the train brakes produce a uniform retardation not exceeding 0.25 m/s²?

Applying Equation (3.4):

$$\bar{V} = 72\,000/3600 \qquad (\text{m/h}) \div (\text{s/h}) = \text{m/s}$$

$$= 20 \text{ m/s}$$

$$2as = \bar{V}^2 - \bar{V}_0^2$$

$$s = \frac{(0)^2 - (20)^2}{2(-0.25)} = 800 \text{ m} = 0.8 \text{ km} \qquad \frac{(\text{m/s})^2}{\text{m/s}^2} = \text{m}$$

Example 3-4

A car is moving at 90 km/h when the brakes are locked and it skids to a halt with constant deceleration. If it takes three seconds to slow down to 45 km/h, at what rate is the car being decelerated; how long does it take to come to a halt; and how far does travel continue before the vehicle is brought to a complete stop?

$$\bar{V}_0 = 90\,000/3600 = 25 \text{ m/s}$$

hence, after three seconds, $\bar{V}_3 = 12.5$ m/s, and the deceleration may be obtained from:

$$\bar{V}_3 = \bar{V}_0 + at$$

$$12.5 = 25 + a(3)$$

$$a = \frac{12.5 - 25}{3} \qquad \frac{\text{m/s} - \text{m/s}}{\text{s}} = \frac{\text{m}}{\text{s}^2}$$

$$= -4.17 \text{ m/s}^2$$

where the time to stop is calculated by:

$$\bar{V} = \bar{V}_0 + at$$

$$t = \frac{0 - 25}{-4.17} = 6 \text{ s} \qquad \frac{\text{m/s} - \text{m/s}}{\text{m/s}^2} = \text{s}$$

and applying Equation (3.5), the distance travelled is:

$$s = \tfrac{1}{2}(\bar{V} + \bar{V}_0)t$$

$$s = \tfrac{1}{2}(0 + 25)6 = 75 \text{ m} \qquad (\text{m/s} + \text{m/s})\text{s} = \text{m}$$

3.6 Free-Falling Bodies

Free-falling bodies also follow the laws of rectilinear motion previously stated but are subject to gravitational attraction which produces an average downward acceleration generally denoted by:

$g = 9.8$ m/s² (approximate average value on earth's surface).

Such influence adds vectorially to any other motion imparted to the body but may be treated as a separate component in computations.

Example 3-5

How long will it take a free body starting from rest, neglecting air friction, to drop 100 meters? What will its velocity be at that point?

$$s = \bar{V}_0 t + \tfrac{1}{2}gt^2$$

$$100 = 0 + \tfrac{1}{2}(9.8)t^2 \qquad \frac{\text{m}}{\text{s}^2} \cdot \text{s}^2 = \text{m}$$

$$t = 4.52 \text{ s} \qquad \frac{\text{m}}{\text{s}} + \frac{\text{m}}{\text{s}^2} \cdot \text{s} = \text{m/s}$$

$$\bar{V} = \bar{V}_0 + gt$$

$$\bar{V} = 0 + 9.8(4.52) = 44.3 \text{ m/s}$$

Example 3-6

A baseball player hits a ball so that it starts from a point 1.30 meters above a level field and moves upward at an initial speed of 35 m/s along a path taking an initial angle of 30 degrees with the horizontal. A fielder

Ch. 3   DYNAMICS   65

catches the ball 1.70 meters above the same datum. Neglecting air friction, what is the apparent distance between the two men?

Assuming the customary set of axes, x horizontal and y vertical:

$$\bar{V}_y = \bar{V}_0 \cdot \sin\theta = 35(0.500) = 17.5 \text{ m/s}$$

$$\bar{V}_x = \bar{V}_0 \cdot \cos\theta = 35(0.866) = 30.3 \text{ m/s}$$

$$s_y = \bar{V}_y \cdot t + \tfrac{1}{2}gt^2$$

$$(1.70 - 1.30) = 17.5t + \tfrac{1}{2}(-9.8)t^2 \qquad \frac{\text{m}}{\text{s}} \cdot \text{s} + \frac{\text{m}}{\text{s}^2} \cdot \text{s}^2 = \text{m}$$

$$t = 3.55 \text{ s}$$

$$s_x = \bar{V}_0 t = 30.3(3.55) = 107 \text{ m}$$

3.7 Effect of Unbalanced Forces

At this point a further explicit review of some first principles of mechanics may again be advisable. As mentioned previously, motion of a body, or change in motion, is due to unbalanced forces which have acted, or are acting on that body.

In assessing the net effect of the "unbalanced" forces it will be essential in various instances to determine the following information for each force:

1. Point of application
2. Orientation of line of action
3. Direction, or sense, along given line of action
4. Magnitude.

Where a number of forces are involved it is generally advisable to sum up algebraically the components along a selected axis or set of axes.

Unbalanced forces may be placed to cause either translation or rotation, or both. The two net effects must be treated separately as will be indicated in the examples which follow.

Example 3-7

A mass of 1.2 metric tons is moved on a horizontal plane by a horizontal force of 8 kN; the motion is resisted by a frictional force of 1750 N. What is the resultant acceleration?

Remembering that for linear motion: $F = ma$

$$8000 - 1750 = 1200a \qquad N = kg \cdot \frac{m}{s^2}$$

$$a = 5.21 \text{ m/s}^2 \qquad N/kg = m/s^2$$

Example 3-8

A suspended mass of 550 grams is raised by a vertical string applying a tension of 8.0 N. What is the net acceleration?

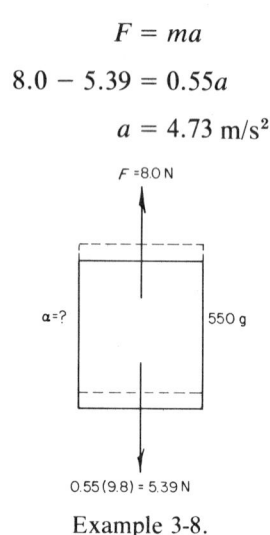

$$F = ma$$

$$8.0 - 5.39 = 0.55a$$

$$a = 4.73 \text{ m/s}^2$$

```
          F = 8.0 N
              ↑
      ┌───────┼───────┐
      │       │       │
      │       │       │
  a=? │       │       │ 550 g
      │       │       │
      │       │       │
      └───────┼───────┘
              ↓
       0.55(9.8) = 5.39 N
```

Example 3-8.

3.8 Impulse and Momentum

Impulse and momentum are both vector quantities. Impulse is the product of a force and the time the force acts. Momentum of a body is a product of the mass of the body and the velocity at which it is moving, thus:

$$\text{Impulse} = Ft \qquad N \cdot s$$

$$\text{Momentum} = m\bar{V} \qquad kg \cdot m/s$$

The total momentum of any system of masses is constant until the system is acted upon by external force. Then the change in momentum is equal to the impulse, thus:

(Eq. 3.6) $\quad Ft = m(\bar{V} - \bar{V}_0) \qquad N \cdot s = kg \cdot \dfrac{m}{s^2} \cdot s = kg \cdot m/s$

Example 3-9

A 3-kg ball, rolling freely at 15 m/s, runs into a stop from which it rebounds at 12 m/s after an impact interval of 10 milliseconds. What force was exerted on the stop?

Applying Equation (3.6):

$$Ft = m(\bar{V} - \bar{V}_0)$$

$$F(0.010) = 3(12 - \{-15\})$$

$$F = 8100 \text{ N} = 8.1 \text{ kN}$$

Example 3-10

A block weighing ½ metric ton (½ Mg) is pulled up a slope of 3.5 percent by a constant force of 4.0 kN. The coefficient of friction between the body and the plane is 0.5. How fast will the block be moving 8 seconds after the force is applied?

$$P_N = 500(9.8)0.999 = 4900 \text{ N}$$

$$P_T = 500(9.8)0.035 = 171 \text{ N}$$

net force along plane is: $4000 - 4900(0.5) - 171 = 1380$ N

$$F \cdot t = m(\bar{V} - \bar{V}_0)$$

$$1380(8) = 500(\bar{V} - 0)$$

$$\bar{V} = \frac{1380}{500}(8) = 22.1 \text{ m/s} \qquad \text{kg} \cdot \frac{\text{m}}{\text{s}^2}(\text{s})\frac{1}{\text{kg}} = \text{m/s}$$

Example 3-10.

3.9 Simple Harmonic Motion

If the displacement (δ) of an oscillating point is measured from the center of its path and the acceleration of the point is proportional and opposite in direction to the displacement, then the motion is called "simple harmonic." It is an important form of motion since it occurs in vibrating springs and in a swinging simple pendulum. This motion can be related to the motion of an analogous object moving at constant speed, $v = r\omega$, on a circular path having a radius equal to the maximum displacement. The projection of the object on a diameter of the circle exhibits simple harmonic motion.

In this analogy the period for one complete cycle, t, of the reference point is the same as that for one revolution of the object on the circular path; namely, $t = 2\pi/\omega$, in which (rad/cycle) ÷ (rad/s) gives $t =$ s/cycle.

Since the linear acceleration of the reference point on the diameter equals $-\delta\omega^2$, it follows that $t = 2\pi(-\delta/a)^{1/2}$ for simple harmonic motion.

For an oscillating spring, $a = k\delta/m$, where k is the spring modulus or stiffness and m is the mass oscillated. Thus, in the same analogous sense the period of oscillation is given by:

(Eq. 3.7) $\qquad t = 2\pi\sqrt{m/k} \qquad \dfrac{\text{rad}}{\text{r}} \cdot \dfrac{\text{s}}{\text{rad}} = \dfrac{\text{s}}{\text{r}}$

Example 3-11

A mass of 800 g is suspended by a spring which has a stiffness factor of 20 N/m. How far is the spring extended in the equilibrium state? If the mass is displaced from the equilibrium position, what would be the frequency of oscillation until equilibrium is restored?

$\qquad$ Force on spring $= 0.800(9.8)$

$\qquad\qquad\qquad\qquad = 7.84$ N $\qquad\qquad$ kg·m/s^2 = N

$\qquad\qquad \delta = 7.84/20$

$\qquad\qquad\quad = 0.392$ m $= 392$ mm $\qquad$ N ÷ N/m = m

Note: Spring constants may also be given in N/mm.

Using Equation (3.7) to calculate the period of oscillation:

$$t = 2\pi\sqrt{\dfrac{0.8}{20}} = 1.257 \text{ s}$$

$\qquad$ Frequency $= 1/t = 1/1.257 = 0.796$ cycle/s $= 0.8$ Hz

Ch. 3   DYNAMICS   69

The SI unit for frequency, or cycles per second, is the "hertz" (Hz) which is 1/s.

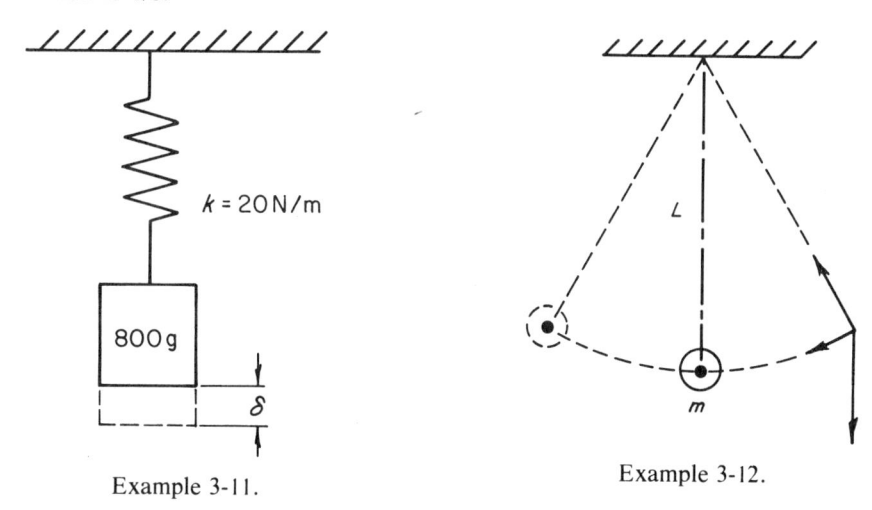

Example 3-11.

Example 3-12.

3.10 SI Units of Empirical Factors

Many equations of mechanics include empirical factors such as k which, being determined from experiment, also require proper interpretation from a units point of view. Thus it is essential to express these "factors" always in terms of the unit system being used. For instance, assuming that 2π incorporates the proper units as previously shown in the circular analogy, the required units for k in Equation (3.7) may be arrived at as follows:

$$t = 2\pi\sqrt{m/k} \qquad t^2 \sim m/k \qquad k \sim m/t^2$$

but,

$$m/t^2 \sim kg/s^2 = N/m$$

which indicates the requirement for k to be in N/m.

See also similar discussions in Chapter 6, "Fluid Mechanics," especially Section 6.26, "Weirs and Open Channels."

3.11 Simple Pendulum

A similar study of the forces acting on a simple pendulum oscillating at small amplitudes shows that where L is the distance between the point of suspension and the center of mass, and g is the local acceleration of gravity, the period is:

(Eq. 3.8) $\qquad t = 2\pi\sqrt{L/g} \qquad \dfrac{\text{rad}}{r}\left(m \cdot \dfrac{s^2}{m}\right)^{1/2} = \dfrac{s}{r}$

Example 3-12

What is the period of oscillation of a mass of 0.8 kg suspended by a cord 0.7 m long?

$$t = 2\pi\sqrt{0.70/9.8} = 1.68 \text{ s}$$

Note—The "period" of oscillation is the total time taken to move from one reference point completely through all other positions and return to the point of origin.

3.12 Work and Energy

The concepts of work and energy involve force and distance, both of which have been discussed. The appropriate SI units are derived from the newton and the meter.

Work is the vector product of force and the distance of movement along the line of action of that force. The SI unit of work is therefore N·m which, as a Derived Unit, has been given the special name of "joule" (J).

The unit "joule" may be utilized only where actual movement occurs in the direction of the force being considered and where actual work has thus been performed. It is *not* proper to use the joule for static moment (force times perpendicular moment arm). Even though the same product (N·m) is perceived in terms of units, the statement of static moment is no more than a unit-product and must be treated solely as such.

Energy is the result of work or the potential to do work. To summarize in general notation:

(Eq. 3.9) $\qquad W = Fs \qquad (\text{kg·m/s}^2)\text{m} = \text{N·m} = \text{J}$

The foregoing applies to linear motion. The "joule" is also used, of course, in describing rotary motion; see 3.21, "Energy of Rotation." In addition, the joule is used as the unit of energy for all electrical, fluid, and thermal phenomena; see Chapters 5, 6, 7, and 8.

Ch. 3 DYNAMICS 71

3.13 Potential Energy

Potential energy arises from the position of a body in a field of force or from elastic deformation. For instance, a body of mass m in a location where the acceleration due to gravity is g, if raised to a height h above a given datum, is said to have a potential energy in respect to that datum, which is stated in SI units as follows:

(Eq. 3.10) $\quad E_p = mgh \quad \text{kg(m/s}^2\text{)m} = \text{kg·m}^2/\text{s}^2 = \text{N·m} = \text{J}$

Example 3-13

A metal casting having a mass of 150 kg is lifted a distance of 8 m. What potential energy does this casting have in respect to its original position?

$$E_p = 150(9.8)8 = 11\,800 \text{ J} = 11.8 \text{ kJ} \quad \text{kg} \cdot \frac{\text{m}}{\text{s}^2} \cdot \text{m} = \text{N·m} = \text{J}$$

3.14 Strain Energy

The behavior of a mechanical spring is generally measured in terms of a spring constant, k, which represents the ratio of applied force to resultant deflection. The SI units of k are preferably N/m. If the spring is compressed at a constant rate to a deflection, δ, then the average force applied is ½ the maximum force. The work done to the spring and the potential energy accompanying the process will be given by the expression:

(Eq. 3.11) $\quad E_p = \tfrac{1}{2}k\delta^2 \quad \dfrac{\text{N}}{\text{m}} \cdot \text{m}^2 = \text{N·m}$

Example 3-14

A coil spring having a spring constant of 120 kN/mm (120 MN/m) is deflected 6.0 mm. What potential energy is thus stored in this spring? By Equation (3.11):

$$E_p = \tfrac{1}{2}(120)(6)^2 = 2160 \text{ J} \quad \dfrac{\text{kN}}{\text{mm}} \cdot \text{mm}^2 = \text{N·m} = \text{J}$$

3.15 Kinetic Energy

Kinetic energy of a body is measured in terms of its motion. When a mass m, starting from rest, achieves a linear velocity $\bar{V}$, it will have

moved a distance s. The work done may be expressed in terms of applied force and distance moved as follows:
Since
$$F = ma$$
and
$$2as = \bar{V}^2 - \bar{V}_0^2$$
and since
$$\bar{V}_0 = 0$$

$$\text{Work} = F \cdot s = (ma)\frac{\bar{V}^2}{2a} = \tfrac{1}{2}m\bar{V}^2$$

(Eq. 3.12) $\quad E_k = \tfrac{1}{2}m\bar{V}^2 \qquad \text{kg}\cdot\text{m}^2/\text{s}^2 = \text{N}\cdot\text{m} = \text{J}$

Example 3-15

What is the amount of kinetic energy which must be absorbed or dissipated to bring a 1-metric-ton vehicle that is traveling at 80 km/h to a complete stop?

By Equation (3.12):

$m = 1.0\text{ t} = 1.0\text{ Mg} = 1000\text{ kg}$

$\bar{V} = 80\text{ km/h} = 80\,000/3600 = 22.2\text{ m/s}$

$E_k = \tfrac{1}{2}(1000)(22.2)^2 = 246\,420\text{ J}$

$ = 246\text{ kJ} \qquad\qquad\qquad (\text{kg})\dfrac{\text{m}^2}{\text{s}^2} = \text{N}\cdot\text{m} = \text{J}$

Example 3-16

A mass of 200 kg is being moved on a horizontal conveyor belt at a speed of 3 m/s when the belt is brought to an abrupt stop. How far will the body slide along the belt if the coefficient of friction is 0.2?

Vertical force on belt = (200)9.8 = 1960 N

Horizontal friction force available = 0.2(1960) = 392 N

Applying Equation (3.12):

$E_k = \tfrac{1}{2}m\bar{V}^2 = \tfrac{1}{2}(200)(3)^2 = 900\text{ J} = 900\text{ N}\cdot\text{m}$

$900/392 = 2.30\text{ m} \qquad\qquad\qquad \text{N}\cdot\text{m/N} = \text{m}$

Ch. 3　　　　　　　　　　DYNAMICS　　　　　　　　　　73

Note again, the positive characteristic of SI, in that *g* appears where it *is* involved with the application of vertical force to the conveyor belt and the resultant friction, but *g* does *not* appear in the expression for kinetic energy of the body moving in a horizontal plane.

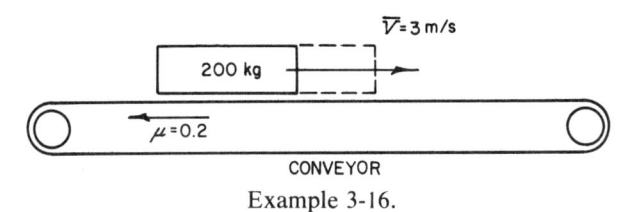

Example 3-16.

In some situations it is advantageous to develop a direct relationship between force and work as demonstrated in Example 3-17.

Example 3-17

A pile driver of 100 kg mass falls 10 meters and impacts on a pile which moves downward 0.3 m. What is the average force exerted? Applying Equation (3.9):

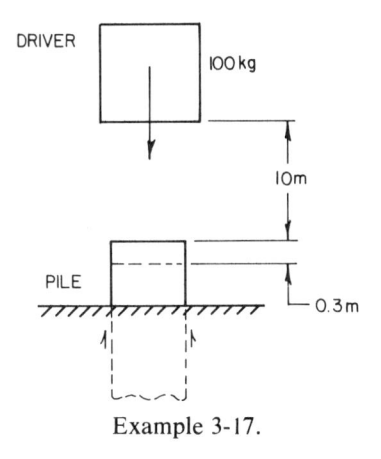

Example 3-17.

$$\text{Force} = 100 \times 9.8 = 980 \text{ N}$$
$$\text{Work} = Fs = 980 \times 10.3 = 10\,094 \text{ J} \quad \text{N·m}$$
$$10\,094/0.3 = 33\,650 \text{ N}, \quad \text{or,} \quad 34 \text{ kN}$$

Note once again, that in the absolute terms of SI, consistency is obtained in the statement that impulse equals change in momentum.

Interrelationships between potential energy and kinetic energy, as well as between momentum and kinetic energy are illustrated in Example 3-18.

Example 3-18

A 50-gram pellet is fired into a stationary 2-kg block which is suspended on a thin wire. As a result of the impact the combined mass swings to a point 150 mm above the original position of the block. What was the original velocity of the pellet?

Calculate first the indicated velocity of the combined mass immediately after impact by equating kinetic energy to the observed potential energy, namely:

Using Equations (3.12) and (3.10)

$$\tfrac{1}{2}m\bar{V}^2 = mgh$$

$$\bar{V} = \{2(9.8)0.150\}^{1/2} \qquad \left\{\frac{m}{s^2}\cdot m\right\}^{1/2} = m/s$$

$$\bar{V} = 1.72 \text{ m/s}$$

Since momentum before impact equals momentum after impact as expressed in Equation (3.6):

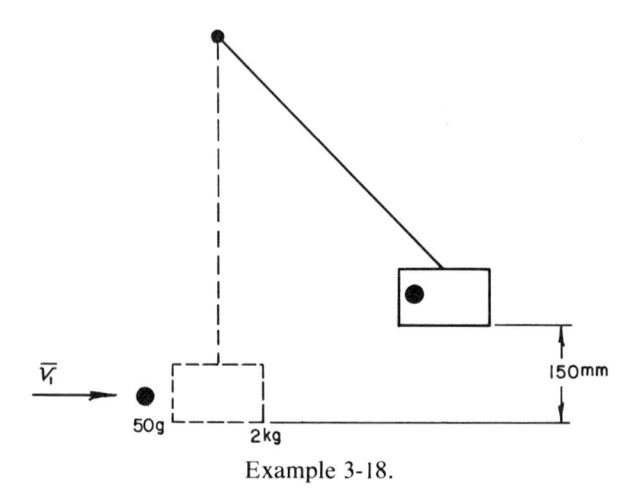

$$m_1\bar{V}_1 = (m_1 + m_2)\bar{V}_2$$

$$50\bar{V}_1 = (50 + 2000)1.72 \qquad (\text{kg}\cdot\text{m/s}) \div \text{kg} = \text{m/s}$$

$$\bar{V}_1 = 70.5 \text{ m/s}$$

Example 3-18.

3.16 Power

Power is the energy generated or consumed per unit of time; it is also work or quantity of heat per unit of time. In terms of "quantity" symbols power may be expressed as:

(Eq. 3.13) $\quad P = Fs/t \qquad kg \cdot \dfrac{m}{s^2} \cdot m \cdot \dfrac{1}{s} = \dfrac{N \cdot m}{s} = \dfrac{J}{s} = W$

It is highly significant that the unit for all power, including both mechanical and electrical as well as the unit for rate of heat transfer, is the watt (W), which is defined as: $W = J/s$. From this it can be seen that work or energy may also be measured by the watt-second, i.e., $J = W \cdot s$.

Power may also be described in terms of Base Units as follows:

$$W = J/s = (kg \cdot m^2/s^2) \div s = kg \cdot m^2/s^3$$

Example 3-19

A metal casting having a mass of 150 kg (see Example 3-13) is to be lifted a distance of 8 m in 16 seconds by means of a crane. What is the power required?

$\quad 150(9.8)8 = 11\,760 \text{ J} \qquad 11\,760/16 = 735 \text{ W} = 0.735 \text{ kW}$

3.17 Rotational Motion

Rotation about a center or an axis introduces the concepts of angular velocity (ω), and angular acceleration (α). To describe rotational motion requires use of the unit of angular measure which in SI is basically the radian (rad).

The radian is used in dynamics because of its coherency. In a circle, the radii forming an angle of one radian at the center intercept an arc on the circumference equal in length to the radius. Thus, in the equation for arc in terms of radius, $s = kr\theta$, if θ is 1.0 radian, then s and r are equal and k, the proportionality factor, is 1/rad. Because the radian is a coherent unit there is a tendency to consider it as a pure number. Angle is *not* a dimensionless quantity as is readily evident if a non-coherent unit such as the degree is used. If such were the case, a proportionality factor of $2\pi/360$ (approximately 0.01745/degree) must be introduced.

The use of the radian measure in rotational analyses is necessary, of course, in order to properly relate the motion of particles which are at different distances from the center of rotation. In this connection it is essential to recall that one revolution equals 2π radians which, in turn, equals 360 degrees. From this it also may be recalled that one r/min equals $2\pi/60 = 0.1047$ rad/s. Obviously, one r/s equals 2π rad/s.

The principles and expressions for rotational motion are similar to those for rectilinear motion except for the distinctive notation and units used, thus:

(Eq. 3.14) $\quad \omega = \omega_0 + \alpha t \qquad\qquad \omega_0 =$ initial angular velocity $\quad$ rad/s

(Eq. 3.15) $\quad \theta = \omega_0 t + \frac{1}{2}\alpha t^2 \qquad \omega =$ final angular velocity $\quad$ rad/s

(Eq. 3.16) $\quad 2\alpha\theta = \omega^2 - \omega_0^2 \qquad\quad \alpha =$ angular acceleration $\quad$ rad/s^2

$\qquad\qquad\qquad\qquad\qquad\qquad\quad \theta =$ angular displacement $\quad$ rad

$\qquad\qquad\qquad\qquad\qquad\qquad\quad t =$ elapsed time $\qquad\qquad\quad$ s

Example 3-20

A flywheel rotating at 102 r/min is speeded up to 120 r/min during an operation which takes ¾ second. What acceleration has been imparted and what is the angular motion in this period?

Applying Equations (3.14) and (3.15):

$\omega = \omega_0 + \alpha t$

$\alpha = (120 - 102)\dfrac{2\pi}{60} \cdot \dfrac{1}{0.75} \qquad\qquad \dfrac{r}{min} \cdot \dfrac{rad}{r} \cdot \dfrac{min}{s} \cdot \dfrac{1}{s} =$ rad/s^2

$\alpha = 2.51$ rad/s^2

$\theta = \omega_0 t + \frac{1}{2}\alpha t^2 \qquad\qquad \omega_0 = (102)2\pi/60 = 10.68$ rad/s

$\theta = 10.68(0.75) + \frac{1}{2}(2.51)(0.75)^2 \qquad\qquad \dfrac{rad}{s}(s) + \dfrac{rad}{s^2}(s)^2 =$ rad

$\theta = 8.72$ rad

$\qquad\qquad 8.72/2\pi = 1.39$ revolutions

3.18 Circular Motion; Centrifugal Action

The tendency of any mass, m, moving at a constant velocity, $\bar{V}$, is to continue to move in a straight line. When circular motion is desired, as shown in Fig. 3-1, this means that the mass particle must move away from the original tangent. To produce constant circular motion there must be introduced a constant radial acceleration. To have a radial

Ch. 3   DYNAMICS   77

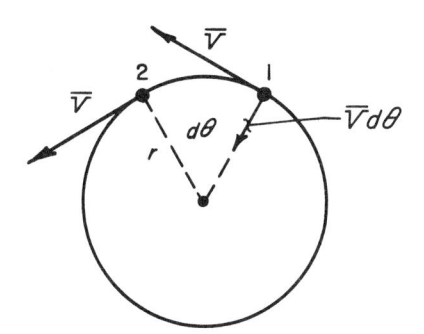

Fig. 3-1. Circular Motion; Centrifugal Action.

acceleration requires the presence of a radial force, generally termed the centripetal force, F_c.

The analysis of circular motion involves products of vectors which must be handled with due consideration of their respective directions. The radial force mentioned above, may be evaluated by an analysis such as the following, in which $\bar{V}$ and $\underline{r}$ are perpendicular to the direction of motion and therefore carry with them the unit characteristic 1/rad for dimensional integrity.

The effective acceleration is:

$$a = \bar{V}\omega \qquad \frac{m}{s \cdot rad} \cdot \frac{rad}{s} = m/s^2$$

Also:

(Eq. 3.17) $\qquad \bar{V} = \underline{r}\omega \qquad \frac{m}{rad} \cdot \frac{rad}{s} = m/s$

Combining the above, yields:

$$a = \bar{V} \cdot \underline{\bar{V}}/\underline{r} = \bar{V}^2/r \qquad \frac{m}{s} \cdot \frac{m}{s \cdot rad} \cdot \frac{rad}{m} = m/s^2$$

and

$$a = \omega\underline{r}\omega = \omega^2\underline{r} \qquad \frac{rad^2}{s^2} \cdot \frac{m}{rad^2} = m/s^2$$

Since $F = ma$, it follows that:

(Eq. 3.18) $\qquad F_c = m\underline{r}\omega^2 \qquad kg \cdot \frac{m}{rad^2} \cdot \frac{rad^2}{s^2} = kg \cdot \frac{m}{s^2} = N$

and

(Eq. 3.19) $\quad F_c = m\bar{V}^2/r \qquad\qquad \text{kg} \cdot \dfrac{m^2}{s^2} \cdot \dfrac{1}{m} = \text{kg} \cdot \dfrac{m}{s^2} = N$

Additional discussion of the applicable units where the phenomena of torque, rotational inertia, work, and energy are involved, will be found in Section 3-19, "Torque and Inertia" and subsequent examples.

Example 3-21

A vehicle having a mass of 2.6 Mg (2.6 metric ton) moves around a circular track of 150 m radius at a speed of 90 km/h. What inward slope of the track is desirable to offset any tendency of the vehicle to overturn? By Equation (3.19):

$F_c = m\bar{V}^2/r$

$F_c = 2600(25)^2/150 = 10\,830 \text{ N} \qquad \text{kg} \cdot \dfrac{m^2}{s^2} \cdot \dfrac{1}{m} = \text{kg} \cdot m/s^2 = N$

$G = 2600(9.8) = 25\,500 \text{ N} \qquad\qquad \text{kg} \cdot \dfrac{m}{s^2} = N$

$\bar{V} = 90\,000/3600 = 25 \text{ m/s}$

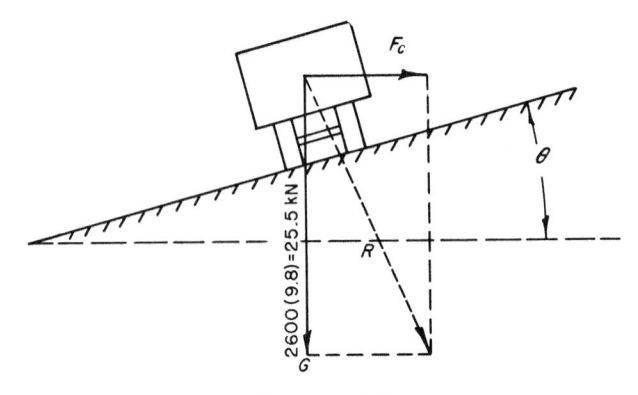

Example 3-21.

to make R normal to vehicle track:

$\tan \theta = 10\,830/25\,500 = 0.425$

$\theta = 25.1°, \quad \text{or,} \quad 4.25/10 \text{ m/m}$

Ch. 3   DYNAMICS   79

Example 3-22

A 4-kilogram mass rolls freely on a horizontal surface maintained by a light wire in a circular path 4.4 meters in diameter at 36 revolutions per minute. What is the tension in the wire?

$$\omega = 2\pi(36/60) = 3.77 \text{ rad/s}$$

Using Equation (3.17):

$$\bar{V} = \omega\, \underline{r} = 3.77(2.2/1.0) = 8.29 \text{ m/s} \qquad \frac{\text{rad}}{\text{s}} \cdot \frac{\text{m}}{\text{rad}} = \frac{\text{m}}{\text{s}}$$

Applying Equation (3.19), the tension in the wire is:

$$F_c = m\bar{V}^2/r \qquad F_c = 4.0(8.29)^2/2.2 = 125 \text{ N},$$

or Equation (3.18):

$$F_c = m\underline{r}\omega^2 = 4.0[2.2/(1.0)^2](3.77)^2 = 125 \text{ N}$$

$$\text{kg} \cdot \frac{\text{m}}{\text{rad}^2} \cdot \frac{\text{rad}^2}{\text{s}^2} = \text{kg} \cdot \frac{\text{m}}{\text{s}^2} = \text{N}$$

In the above, note that $\underline{r} = (2.2/1.0)$ m/rad, and that $\underline{r} = 2.2/(1.0)^2$ m/rad².

Example 3-23

A body on the end of a rope 1.5-m long is whirled in a vertical circle at a speed of 95 r/min. If the body is released when it makes an angle of 30° above the horizontal, how high will the body rise?

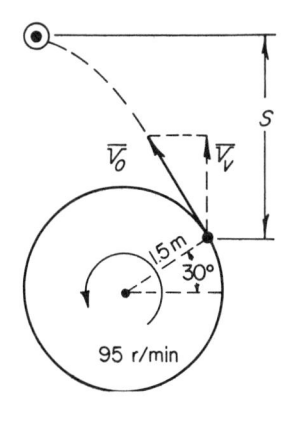

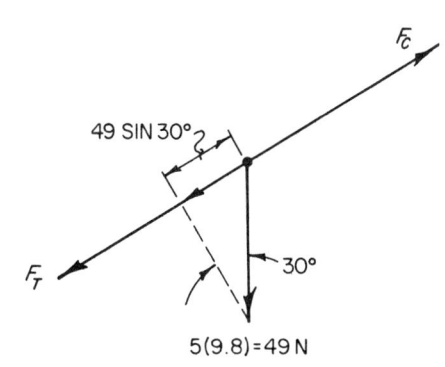

Example 3-23.

80 DYNAMICS Ch. 3

$$\omega = 95(2\pi/60)$$

$$\frac{r}{\min} \cdot \frac{\text{rad}}{r} \cdot \frac{\min}{s} = \frac{\text{rad}}{s}$$

$$= 9.95 \text{ rad/s}$$

$$\bar{V}_0 = 9.95(1.5)$$

$$\frac{\text{rad}}{s} \cdot \frac{m}{\text{rad}} = \frac{m}{s}$$

$$= 14.92 \text{ m/s}$$

or,

$$\bar{V}_0 = \underline{r}\omega = 1.5(95)2\pi/60$$

$$\frac{m}{\text{rad}} \cdot \frac{r}{\min} \cdot \frac{\text{rad}}{r} \cdot \frac{\min}{s} = \frac{m}{s}$$

$$= 14.92 \text{ m/s}$$

Vertical component of initial velocity is:

$$\bar{V}_v = 14.92(0.866) = 12.92 \text{ m/s}$$

Since

$$2gs = \bar{V}^2 - \bar{V}_0^2 \qquad \frac{m}{s^2} \cdot m = \frac{m^2}{s^2}$$

$$-2(9.8)s = (0)^2 - (12.92)^2$$

$$s = 8.52 \text{ m}$$

What is the tension in the rope just before the body is released, if the mass of the body is 5.0 kg?

$$5.0(9.8) = 49 \text{ N} \qquad (49) \sin 30° = 24.5 \text{ N}$$

By Equation (3.18):

$$F_c = m\underline{r}\omega^2 = 5.0(1.5)(9.95)^2 = 742 \text{ N} \qquad \text{kg} \cdot \frac{m}{\text{rad}^2} \cdot \frac{\text{rad}^2}{s^2} = \text{kg} \cdot \frac{m}{s^2} = \text{N}$$

or, by Equation (3.19):

$$F_c = m\bar{V}^2/r = 5.0(14.92)^2/1.5 = 742 \text{ N} \qquad \text{kg} \cdot \frac{m^2}{s^2} \cdot \frac{1}{m} = \text{kg} \cdot \frac{m}{s^2} = \text{N}$$

thus, the tension in the rope is:

$$F_t = 742 - 24.5 = 718 \text{ N}$$

This example clearly illustrates the SI attitude toward "g," discussed earlier, namely that gravity enters the calculations only where it is relevant. Thus, the superior logic of the SI procedure is evident.

Two other points deserve mention:

(1) Data on angular measurements are provided in this example, *not* in terms of the standard SI unit—the radian—but rather in terms of two alternate units, namely: degrees and revolutions;
(2) A link is introduced between circular and linear motion; i.e., the velocity of the moving mass is the product of the radius of the path of motion in meters per radian, and the angular velocity in radians per second. The resultant peripheral velocity is therefore in m/s.

It is interesting to consider this example expressed entirely in terms of the radian. The original statement would have read: a body at the end of a rope 1.5 m long is whirled in a vertical circle at an angular velocity of 9.95 rad/s. If the body is released when the rope makes an angle of 0.52 rad above the horizontal, how high will the body rise?

The linear velocity is then (9.95) rad/s (1.5) m/rad = 14.92 m/s. The vertical component is, therefore, 14.92 cos 0.52 rad. Tables of trigonometrical functions in terms of the radian are available so that the rest of the work can proceed exactly as before. The whole calculation is in fact slightly more simple, but what emerges very clearly is the difficulty of using the radian for making measurements. One could, of course, use a protractor graduated in radians for measuring angles, but the prospect of estimating the angular velocity of the rope other than by counting complete revolutions is discouraging. The inexorable fact that the ratio of the circumference of a circle to its radius is 2π cannot be escaped.

3.19 Torque and Inertia

Rotary motion is most frequently used to store and to transmit energy. Hence, a key to the understanding of this phenomenon rests largely with the concept of energy. For purposes of review, the parallel analysis of linear motion relationships and rotary motion relationships shown in Table 3-3 will be of value.

From Table 3-3 it will be noted that the name "torque" is properly applied only to the quantity pertaining to dynamics; that is, where motion occurs, hence the units J/rad. The simple product of force times moment arm is more properly termed "torsional moment" and is to be placed in the same classification as "bending moment."

Also in connection with Table 3-3, it is well to review the well-known expression for power and to observe the SI units involved.

$$P = 2\pi n T \qquad \frac{\text{rad}}{\text{r}} \cdot \frac{\text{r}}{\text{s}} \cdot \frac{\text{N} \cdot \text{m}}{\text{rad}} = \frac{\text{N} \cdot \text{m}}{\text{s}} = \frac{\text{J}}{\text{s}} = \text{W}$$

Table 3-3. SI Units for Describing Motion

Quan. Symbol	Quantity	Linear Motion	Rotary Motion
$L; \theta$	amount of motion	m	rad
$\bar{V}; \omega$	velocity	m/s	rad/s
$a; \alpha$	acceleration	m/s^2	rad/s^2
$W; E_k$	energy (N·m)	J	J
$F; T$	source of motion	N = J/m (force)	N·m/rad = J/rad (torque)
$m; \bar{I}$	inertia	kg	kg·m^2/rad^2
$m\bar{V}; \bar{I}\omega$	momentum	kg·m/s	kg·m^2/(rad·s)
$Ft; Tt$	impulse	N·s = kg·m/s	(N·m/rad)s = kg·m^2/(rad·s)

Too often the 2π term is considered incorrectly as a dimensionless constant. As is evident, the 2π term carries the units rad/r, which must be balanced out by $T = $ N·m/rad.

A further analysis of the aspects of rotary motion is shown in Fig. 3-2 which provides another parallel analysis on a geometric basis of the Quantities involved and the applicable Units.

Thus, the appropriate inclusion of the radian in all aspects of rotary motion is essential for a full understanding of the phenomenon and for the correct use of a unit system such as SI, which includes the radian as the reference unit for angular measurement.

Admittedly, there are frequent instances in current engineering practice in which the radian is overlooked or ignored, and the results are nevertheless satisfactory. These particular instances occur where: (a) appraisals of values are relative and the omission of the radian in each part of the appraisal is neutralized; (b) measurements are consistently referred to the periphery of a simple circular element, in terms of the radius of the circle; (c) rotation continues through numerous full revolutions and the radian, as a dimensional aspect of the factor 2π, is overlooked.

Whatever the reason for ignoring the radian in some everyday practice, it will be treated fully in all of the Examples which follow so that the reader may at all times be aware of its proper presence.

Turning next to examples of rotary motion it is well to re-state the definition of quantities involved and the essential relationships as expressed in the key formulas. Remembering that motion is occurring

Ch. 3 DYNAMICS 83

(a) In Terms of Quantities (b) In Terms of Units

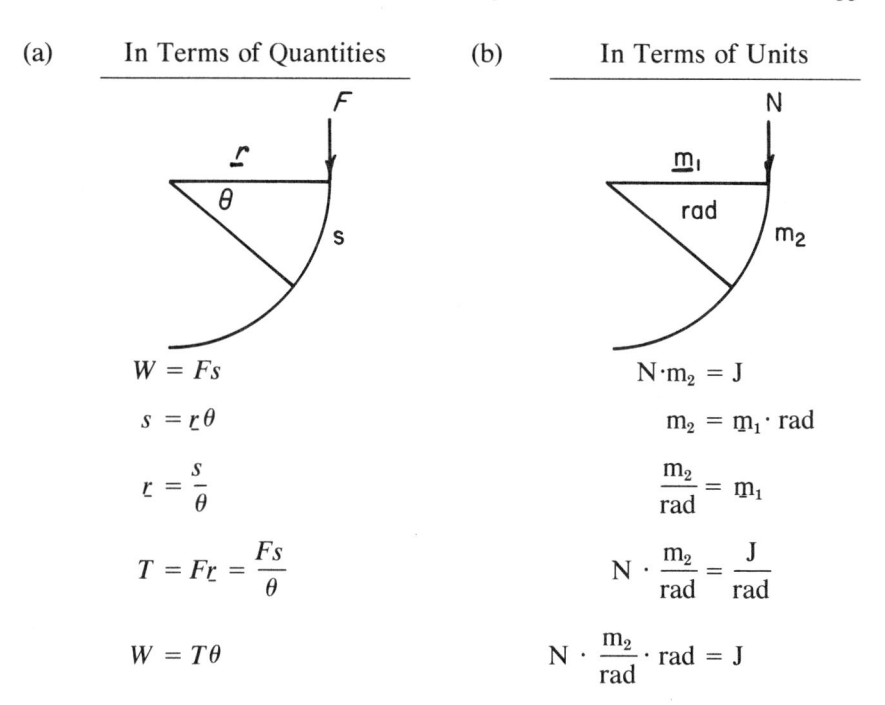

$W = Fs$ $N \cdot m_2 = J$

$s = \underline{r}\theta$ $m_2 = \underline{m}_1 \cdot rad$

$\underline{r} = \dfrac{s}{\theta}$ $\dfrac{m_2}{rad} = \underline{m}_1$

$T = F\underline{r} = \dfrac{Fs}{\theta}$ $N \cdot \dfrac{m_2}{rad} = \dfrac{J}{rad}$

$W = T\theta$ $N \cdot \dfrac{m_2}{rad} \cdot rad = J$

Fig. 3-2. A study of geometric relationships in rotary motion.

and that all of the following is for rotation in a plane about the center of the mass:

T = unbalanced torque causing rotation $N \cdot m/rad$

$\bar{I}$ = inertia of mass about the axis of rotation $kg \cdot m^2/rad^2$

α = angular acceleration rad/s^2

(Eq. 3.20) $T = \bar{I}\alpha$ $\dfrac{kg \cdot m^2}{rad^2} \cdot \dfrac{rad}{s^2} = \dfrac{kg \cdot m^2}{s^2} \cdot \dfrac{1}{rad} = N \cdot m/rad$

If the rotating mass is a solid circular element of uniform density, ρ; uniform thickness, L; and diameter, $2r$; as is frequently the case, the value of inertia is evaluated by using the formula below for which the units are as indicated:

$$\bar{I} \sim \rho(\pi r^2)L(\underline{r}^2/2) \quad \text{but} \quad m = \rho L(\pi r^2),$$

hence:

(Eq. 3.21) $\bar{I} = \frac{1}{2} m\underline{r}^2 \qquad \text{kg} \cdot \frac{m^2}{rad^2}$

It is common practice, in connection with rotating bodies generally, to introduce the concept of a "radius of gyration" symbolized by k. This is interpreted as the distance from the center of rotation to an arc on which the entire mass might be concentrated and have the *same* rotational inertia, thus:

(Eq. 3.22) $\bar{I} \propto mk^2$, where k is usually stated in meters.

Actually, k should preferably be termed "gyrational factor," or, "inertial factor" and be stated in the unit of meter per radian (m/rad).

Example 3-24

A solid flywheel of uniform thickness has a diameter of 1.5 m and a mass of 800 kg. What torque is needed to increase its angular velocity by 100 r/min in 1.2 seconds?

$\bar{I} = \frac{1}{2} m\underline{r}^2 = \frac{1}{2}(800)(0.75/1.0)^2 = 225 \text{ kg} \cdot m^2/rad^2$

change of 100 r/min $= \Delta\omega = 100(0.1047) = 10.47$ rad/s

$\Delta\omega = \alpha t \qquad \alpha = \Delta\omega/t \qquad \alpha = 10.47/1.2 = 8.73 \text{ rad/s}^2$

$T = \bar{I}\alpha \qquad T = 225(8.73) = 1960 \text{ N} \cdot \text{m/rad} \qquad \frac{kg \cdot m^2}{rad^2} \cdot \frac{rad}{s^2} = \frac{N \cdot m}{rad}$

3.20 Angular Momentum and Impulse

In a manner similar to that for linear momentum and impulse, it may be shown that:

(1) angular momentum is the product of rotational inertia and angular velocity:

Angular Momentum $= \bar{I}\omega \quad (\text{kg} \cdot m^2/rad^2)(rad/s) = \text{kg} \cdot m^2/(rad \cdot s)$

(2) angular impulse is the product of torque and the time it acts:

Angular Impulse $= Tt \quad (N \cdot m/rad)s = \text{kg} \cdot m^2/(rad \cdot s)$

Thus:

(Eq. 3.23) $Tt = \bar{I}(\omega - \omega_0)$

Example 3-25

A flywheel with a rotational inertia of 20 kg·m²/rad² is revolving with an angular velocity of 8 rad/s when a constant torque of 30 N·m/rad is applied in opposition to its rotation. How long must this torque act to reverse the flywheel and accelerate it to a speed of 7 rad/s in the opposite direction?

Change in angular momentum:

$$\bar{I}(\omega - \omega_0) = 20[-7 - (8)] = -300 \text{ kg·m}^2/(\text{rad·s})$$

due to impulse Tt, whence, by Equation (3.23):

$$t = \bar{I}(\omega - \omega_0)/T \qquad t = 300/30 = 10.0 \text{ s} \qquad \frac{\text{kg·m}^2}{\text{rad}^2} \cdot \frac{\text{rad}}{\text{s}} \cdot \frac{\text{s}^2 \cdot \text{rad}}{\text{kg·m}^2} = \text{s}$$

3.21 Energy of Rotation

For rotational movement the work done by a constant torque is measured in terms of $T\theta$:

(Eq. 3.24) $\qquad W = T\theta \qquad \dfrac{\text{N·m}}{\text{rad}} \cdot \text{rad} = \text{N·m} = \text{J}$

Alternately, this may be viewed by combining Equations (3.16) and (3.24) as follows:

Since $\qquad 2\alpha\theta = \omega^2 - \omega_0^2; \quad T = \bar{I}\alpha; \quad \text{and} \quad \theta = E_k/t$

(Eq. 3.25) $\qquad E_k = \tfrac{1}{2}\bar{I}(\omega^2 - \omega_0^2) \qquad \dfrac{\text{kg·m}^2}{\text{rad}^2} \cdot \dfrac{\text{rad}^2}{\text{s}^2} = \text{kg·m}^2/\text{s}^2 = \text{N·m} = \text{J}$

Example 3-26

For the flywheel in Example 3-24, what is the increase in kinetic energy?

Noting that $\omega_0 = 0$ and applying Equation (3.25):

$$E_k = \tfrac{1}{2} \bar{I}(\omega^2 - \omega_0^2) = \tfrac{1}{2} (225)\{(10.47)^2 - (0)^2\} = 12\,300 \text{ J}$$

Where a body has linear motion and is rotating at the same time, the respective E_k values must be computed separately but the final results can be added since both are measured in identical units: N·m = J.

Example 3-27

What power is required to bring about the change in speed of the flywheel cited in Examples 3-24 and 3-26?

$$P = E_k/t = 12\,300/1.2 = 10\,250 \text{ W}$$

or alternately:

(Eq. 3-26) $\quad P = \frac{1}{2}\,T\theta/t$

$$P = \frac{1}{2}(1960)(10.47) \qquad \frac{N \cdot m}{rad} \cdot \frac{rad}{s} = \frac{J}{s} = W$$

$$P = 10\,250 \text{ W} = 10.25 \text{ kW}$$

Example 3-28

The mass of a flywheel is 500 kg and may be assumed as concentrated at 0.6 m from the center of rotation. What energy is given out when its speed is reduced from 1000 r/min to 800 r/min? If this is accomplished in two minutes, what is the average power output?

$$\bar{I} = mk^2 = 500(0.6/1.0)^2 \qquad \bar{I} = 180 \text{ kg} \cdot m^2/rad^2$$

$$\omega_0 = \frac{1000}{60}(2\pi) = 104.7 \text{ rad/s} \qquad \omega = \frac{800}{60}(2\pi) = 83.8 \text{ rad/s}$$

$$E_k = \frac{1}{2}\bar{I}(\omega^2 - \omega_0^2)$$

$$= \frac{180}{2}\{(83.8)^2 - (104.7)^2\} \qquad kg \cdot \frac{m^2}{rad^2}(rad/s)^2 = N \cdot m = J$$

$$E_k = 355\,000 \text{ J} = 355 \text{ kJ} = 0.36 \text{ MJ}$$

$$P = \frac{355\,000}{120} = 2960 \text{ W} = 2.96 \text{ kW}$$

Example 3-29

A mass which has a rotational inertia of 200 kg·m²/rad² is revolving at 30 rad/s. If a braking torque of 50 N·m/rad is applied, what is the angle through which the mass will turn before coming to rest?

Ch. 3 DYNAMICS 87

Applying Equations (3.23), (3.14), and (3.15):

(Eq. 3.23) $Tt = \bar{I}(\omega - \omega_0)$

$50(t) = 200(30 - 0)$

$t = \dfrac{200}{50}(30) \qquad \dfrac{\text{kg} \cdot \text{m}^2}{\text{rad}^2} \cdot \dfrac{\text{rad}}{\text{N} \cdot \text{m}} \cdot \dfrac{\text{rad}}{\text{s}} = \dfrac{\text{kg} \cdot \text{m}^2}{\text{rad}^2} \cdot \dfrac{\text{rad} \cdot \text{s}^2}{\text{kg} \cdot \text{m}^2} \cdot \dfrac{\text{rad}}{\text{s}} = \text{s}$

$t = 120 \text{ s}$

(Eq. 3.14) $\omega = \omega_0 + \alpha t$

$30 = 0 + \alpha(120)$

$\alpha = 30/120 = 0.25 \text{ rad/s}^2 \qquad \dfrac{\text{rad}}{\text{s}} \cdot \dfrac{1}{\text{s}} = \dfrac{\text{rad}}{\text{s}^2}$

(Eq. 3.15) $\theta = \omega_0 t + \tfrac{1}{2}\alpha t^2$

$\theta = 30(120) - \tfrac{1}{2}(0.25)(120)^2 \qquad \dfrac{\text{rad}}{\text{s}} \cdot \text{s} - \dfrac{\text{rad}}{\text{s}^2} \cdot \text{s}^2 = \text{rad}$

$\theta = 3600 - 1800$

$\theta = 1800 \text{ rad}$

4

Strength of Materials

4.1 Stress, Strain, and Elasticity

When assessing the ability of a structure or mechanism to withstand the loads to which it will be subjected, the concepts of stress, strain, and elasticity are introduced in order to make possible a precise determination.

For some Quantity Symbols, SI Units and Formulas frequently used in Strength of Materials see Tables 4-1 and 4-2. For information on preferred SI units and magnitudes of the pertinent factors, in SI, see Table 4-3, "Some Approximate Physical Properties of Representative Engineering Materials." These data are intended for use in this book for illustrative purposes only and are not intended as a reference for design. Note also that Table 4-3 gives the generally preferred SI units for statement of physical values. In a particular problem it may occasionally be found desirable to use an alternate form.

4.2 *Unit Stress* is defined as force per unit area; over a finite area this is an average stress. In SI terms, unit stress is newtons per square meter (N/m^2) and this quantity is given the special name of pascal (Pa). The Base Units of the pascal are ($kg \cdot m/s^2$) $\div$ m^2 which yields $kg/(m \cdot s^2)$. Commonly used multiples of stress units are: kN/m^2, MN/m^2, GN/m^2; or kPa, MPa, and GPa.

4.3 *Unit Strain* is defined as elastic deformation per unit of length and hence is given by the ratio m/m or mm/mm. If the strain is torsional, as in shafts, the unit strain is radians per unit of length: rad/m.

4.4 *Modulus of Elasticity* is that physical property of a material, within its proportional limit, which expresses the relationship between unit stress and unit strain. Thus, unit stress divided by unit strain gives the resultant modulus of elasticity which is in the same units as stress, namely: multiples of N/m^2 or Pa.

In cases of simple tension or short block compression, this modulus is

STRENGTH OF MATERIALS

Table 4-1. Quantities and SI Units Frequently Used in Strength of Materials

Quantity Symbol	Quantity Name	* SI Unit Symbols
A	cross-section area	mm², m²
α	coefficient of thermal expansion	μm/(m·°C)
δ	deflection, elongation	mm
ϵ	unit strain	mm/mm
θ	angle of twist	rad
D, d	diameter	m, mm
d	depth	mm
F	total force, total load	N
f	distributed force	N/m; N/m²
E	modulus of elasticity (linear deformation)	Pa; N/m²
G	modulus of rigidity (shear, torsion)	Pa; N/m²
g	acceleration of gravity	m/s²
I	second moment of area	mm⁴
J	polar second moment of area	mm⁴
L	length	m, mm
m	mass	kg
M	bending moment	N·m; N·mm
M_t	torsional moment	N·m; N·mm
p	unit pressure	kPa; kN/m²
ρ	mass density	kg/m³
r	radius; buckling factor ($r = \sqrt{I/A}$)	mm
$\sigma, \sigma_c, \sigma_t$	unit stress	Pa; N/m²
σ_s	unit shear stress	Pa; N/m²
Δt	temperature interval	°C
t	thickness	mm
V	shear force	N
y	distance to stressed fiber	mm
S	section modulus	mm³

* For complete interpretation of this table, note the following in terms of Base Units:

$$N = kg \cdot m/s^2 \qquad N \cdot m = kg \cdot m^2/s^2 \qquad Pa = N/m^2 = kg/(m \cdot s^2)$$

known as the "stretch modulus," or Young's modulus for which the symbol E is used. In cases of torsion the factor of proportionality is known as the "modulus of rigidity" and is expressed by the symbol G.

4.5 *Simple Stresses* are classified as: tension, compression, bearing, and shear. In many cases they exist alone or can be treated separately; in numerous other cases they exist in combined form and the principal, or maximum and minimum, stresses must be sought. In respect to the

Table 4-2. Formulas Frequently Used in Strength of Materials

Equation Number	Formula	Equation Number	Formula
(4.1)	$\sigma = F/A$	(4.12)	$\sigma = My/I$
(4.2)	$F = p \cdot A$	(4.13)	$\sigma = M/S$
(4.3)	$\delta = FL/AE$	(4.14)	$M = fL^2/8 = FL/8$
(4.4)	$\delta = \sigma L/E$	(4.15)	$\delta = \dfrac{5}{384} \dfrac{fL^4}{EI}$
(4.5)	$\delta = \alpha \Delta t L$		
(4.6)	$\sigma = \delta E/L$		$= \dfrac{5}{384} \dfrac{FL^3}{EI}$
(4.7)	$\sigma = \alpha \Delta t E$	(4.16)	$\sigma_s = V/td$
(4.8)	$\sigma_s = M_t r/J$	(4.17)	$F = \pi^2 EI/CL^2$
(4.9)	$\theta = M_t L/JG$	(4.18)	$F/A = \pi^2 E/C(L/r)^2$
(4.10)	$J = \pi d^4/32$	(4.19)	$r = \sqrt{I/A}$
(4.11)	$J = (d_0^4 - d_i^4)\pi/32$	(4.20)	$\delta = 8\,FND^3/Gd^4$
		(4.21)	$\sigma_s = 8\,FD/\pi d^3$

application of a system of units, such as SI, all of the essential points can be observed by examples in cases of simple stress.

4.6 *Tension* tends to extend or stretch a body and is simple tension if the load is axially applied, i.e., acts through the center of gravity of the section to which it is applied. If so, simple stress analysis generally applies to elements of unlimited length.

A critical aspect of the analysis of tensile stress is to be on the alert for reduced or "net" areas, which are thus critical locations, as in pin or bolt holes, threaded sections, etc.

4.7 *Compression* tends to foreshorten a body or to reduce the axial dimension of a body. Simple compressive stress generally exists only in axially loaded, short compression blocks, defined usually as having length to least dimension ratio equal to or less than 10, or length to gyrational factor of cross section less than 40.

4.8 *Bearing stress* may be viewed as a special form of compression in which the force is applied to an area of contact between two parts, the effective area for which is usually calculated on the basis of the projected area of the actual surface of contact.

4.9 *Shear stress* is the result of opposing forces acting on adjacent planes, representing a cross section or other section of a structural or machine element, and thereby causing angular distortion of the elements at the interface. When the shear forces are applied to pins, bolts, rivets, etc., these elements may be in "single" shear or in "double" shear depending on the arrangement of the connection and the resultant number of planes to which the total force is applied.

4.10 *Thermal stress.* Stress will also result when a material is heated or cooled but is restrained from normal expansion or contraction. The amount of potential "free" (unrestrained) change in length can be determined by utilizing the thermal coefficient of linear expansion for the material which in SI will be expressed in terms of the increment of temperature in degrees Celsius.

Units for the resultant coefficient therefore are: basically m/(m·°C), but due to the small magnitude and the convenience of SI prefixes, are usually stated as μm/(m·°C). The unit restraint can be used directly with the modulus to calculate the induced thermal stress.

4.11 *Stress and Strain Calculations.* The essential procedure in strength-of-materials design work is the calculation of expected stresses and/or strains, followed by comparison with and utilization of, data on the physical properties of the material.

The calculation of static stresses involves only forces, dimensions, and angles, and consequently, involves no factors additional to those considered in Chapter 2. The correct treatment of mass and force must, of course, be observed. Stress concentration factors and safety factors are pure numbers, and are unaffected by a change in units.

Example 4-1

A short strut of 40 mm² cross section is subjected to a compressive load of 5 kN. What is the resulting stress?

(Eq. 4.1) $\sigma = F/A$ $\sigma = 5\,000/(40 \times 10^{-6})$ N/m²

$\sigma = 125 \times 10^6$ N/m²

$\sigma = 125$ MN/m² $= 125$ MPa

92   STRENGTH OF MATERIALS   Ch. 4

Table 4-3. Comparative Analysis of Some Approximate Physical
In Terms of U.S. Customary Units*

Coeff. Linear Expan., α 10^{-6} in./(in.·°F)	Allow. Stresses lb/in.² × 10³			Elastic Modulus, lb/in.² × 10⁶		Weight Density, w lb/ft³
	σ_t	σ_c	σ_s	E	G	
	(a)	(b)				
6.5	20	20	10	30	12	490
6.9	24	24	15	30	12	490
6.0	3	10	2	15	6	450
9.3	8	8	5	17	6.4	560
10.4	12	8	6	13	5	520
13.0	16	15	8	10.3	4	170
1.7	1.3	0.8	0.05	1.2	—	27
2.5	1.8	1.2	0.10	1.6	—	48
6.2	1.2	1.0	0.15	2.5	—	150
—	—	0.03	—	—	—	105
4.4	—	0.3	—	—	—	165
—	—	—	—	—	—	62.4

* Note—Values given are *rounded* in each system and are *not* direct conversions.
** For use *only* for comparing representative values in the respective unit systems; *not* AISC, ACI, IFI, etc. (a) extreme fiber in bending; (b) short compression block; in timber,

There is nothing intrinsically incorrect about writing the expression for stress as: 5 kN/40 mm² = 0.125 kN/mm², but it is generally regarded as poor practice to have two prefixes in the same unit. It is preferred practice to have a single prefix, if any, in the numerator, thus MN/m², or, MPa is recommended.

Example 4-2

A mass of 1.3 kg is suspended from a vertical wire of 1.5 mm diameter. What is the tension in the wire and the resulting stress?

The "weight" is: 1.3(9.8) = 12.7 N. This is the tension. The tensile

Ch. 4 STRENGTH OF MATERIALS 93

Properties of Representative Engineering Materials**

		In Terms of Preferred SI Units*					
Material	Mass Density, ρ kg/m³	Elastic Modulus GN/m² = GPa		Allow. Stresses MN/m² = MPa			Coeff. Linear Expans., α μm/(m·°C)
		E	G	σ_t (a)	σ_c (b)	σ_s	
Mild Steel	7850	200	80	140	140	50	11.7
High Stren. Steel	7850	200	80	165	165	70	12.4
Cast Iron	7200	100	40	20	70	20	10.8
Copper	8960	120	45	55	55	35	16.7
Brass	8300	90	35	80	55	40	18.7
Aluminum	2700	70	27	110	100	55	23.4
Timber							
Softwood	430	9	—	9.6	5.5	0.3	3.1
Hardwood	770	12	—	12.4	8.3	0.7	4.5
Concrete (reinf.)	2400	17	—	8.3	6.9	1.0	11.2
Soil	1680	—	—	—	0.2	—	—
Rock	2640	—	—	—	2.0	—	7.9
Water	1000	—	—	—	—	—	—

intended for design. For design purposes see other standard references such as: ANSI,
parallel to grain.

stress accordingly is:

$$\sigma = F/A$$
$$\sigma = 12.7/\pi(0.75)^2 \times 10^{-6}$$
$$= 7187 \text{ kN/m}^2 = 7.19 \text{ MN/m}^2$$

Example 4-3

A steel plate of cross-section 450 mm × 30 mm is subjected to a direct shearing force of 3 MN. What is the resulting stress?

The shearing stress is:

$$\sigma_s = V/A$$

$$\sigma_s = \frac{3 \times 10^6}{(0.450)(0.030)}$$

$$= 222 \text{ MN/m}^2 = 222 \text{ MPa} \qquad \frac{N}{m \cdot m} = \frac{N}{m^2} = Pa$$

Example 4-4

An inspection port in a pressure vessel is covered by a circular plate of an effective diameter of 150 mm, which is retained by eight bolts of root diameter 20 mm.

What is the tensile stress in the bolts when the pressure in the vessel is 3.0 MPa?

The total force withstood by the bolts is:

(Eq. 4.2) $\quad F = pA$

$F = (3.0 \times 10^6)(\pi/4)(0.150)^2 \qquad \text{Pa} \cdot \text{m}^2 = (N/m^2)m^2 = N$

$F = 53 \text{ kN} = 53\,000 \text{ N}$

The total area of cross section of the bolts is:

$$8(\pi/4)(20)^2 = 2513 \text{ mm}^2$$

The stress is therefore:

$$\sigma = F/A$$

$$\sigma = 53\,000/2513 \times 10^{-6}$$

$$= 21.1 \text{ MN/m}^2 = 21.1 \text{ MPa} \qquad N/m^2 = Pa$$

Example 4-5

A mild steel rod has a length of 1.25 m and a cross-sectional area of 75 mm²; it is subjected to a tension of 3.50 kN. What is the elongation of the rod?

The tensile stress is:

$$\sigma = F/A \quad 3500/75 \times 10^{-6}$$
$$= 46.7 \text{ MN/m}^2 = 46.7 \text{ MPa} \quad \text{N/m}^2 = \text{Pa}$$
$$E = 200 \text{ GPa} \quad L = 1.25 \text{ m}$$

(Eq. 4.3) $\quad \delta = FL/AE \quad \quad \text{N·m} \div \text{m}^2(\text{N/m}^2) = \text{m}$

(Eq. 4.4) $\quad \delta = \sigma L/E \quad \quad (\text{N/m}^2)\text{m} \div (\text{N/m}^2) = \text{m}$

$$\delta = 46.7 \times 10^6 (1.25) \div 200 \times 10^9$$
$$= 0.00029 \text{ m} = 0.29 \text{ mm}$$

or,

$$\delta = 46.7 \times 10^6 (1250) \div 200 \times 10^9 = 0.29 \text{ mm}$$

Example 4-6

A steel rail of length 30 m is constrained at the ends but in a manner that there is no stress at 20°C. What will be the stress at −10°C? The coefficient of expansion is 12 μm/(m·°C), and Young's modulus is 200 GPa.

Observe the use of the degree Celsius (°C) to denote an increment in temperature when defining an ordinary temperature change situation. A degree Celsius (°C) is, of course, equivalent to the kelvin (K) which is the Base Unit to be used when quoting thermodynamic conditions.

(Eq. 4.5) $\quad \delta = \alpha \Delta t L \quad (\text{m})(°\text{C})[\text{m/(m·°C)}] = \text{m}$

$$\Delta t = \{20 - (-10)\}$$
$$\Delta t = 30$$
$$\delta = (12 \times 10^{-6})30(30) = 0.0108 \text{ m} = 10.8 \text{ mm}$$

(Eq. 4.6) $\quad \sigma = \delta E/L \quad \text{m}(\text{N/m}^2)/\text{m} = \text{N/m}^2 = \text{Pa}$

$$\sigma = 0.0108(200)/30 = 0.072 \text{ GPa} = 72 \text{ MPa}$$

or more directly:

(Eq. 4.7) $\quad \sigma = \alpha \Delta t E \quad (°\text{C})(\text{N/m}^2)[\text{m/(m·°C)}] = \text{N/m}^2 = \text{Pa}$

$$\sigma = 12 \times 10^{-6}(30)200 \times 10^9 = 72 \text{ MPa}$$

4.12 Torsion in Shafts

The stresses and elastic displacement produced by simple torsional moment in circular shafts may be analyzed as indicated in Fig. 4-1, by the following expressions:

(Eq. 4.8) $\sigma_s = M_t r/J$ N·mm(mm)/mm^4 = N/mm^2 = MN/m^2 = MPa

(Eq. 4.9) $\theta = M_t L/JG$ N·mm(mm) ÷ (mm)^4·N/mm^2 = mm/mm = rad

in which the terms and usual stated units are:

M_t = torsional moment   N·mm

J = polar second moment of cross-sectional area   mm^4

(Eq. 4.10) $J = \pi d^4/32$, for a solid circular shaft

r = radius to fiber at which stress is determined   mm

σ_s = unit shear stress   N/mm^2 = MN/m^2, or MPa

θ = angle of twist   rad, or mm/mm

L = length subject to twist   mm

G = modulus of rigidity   GN/m^2, or GPa

4.13 Compatible Unit Multiples

All of the above quantities must, of course, be introduced in any computation in compatible form of multiples or sub-multiples of Base Units and/or Derived Units. This fact leads to the observation that in most SI references, as in Table 4-3, G will be stated in accordance with the preferred SI practice, in a multiple best suited to its magnitude; that is, in terms of GN/m^2, or GPa. However, since it is more convenient in this particular computation to have the value of G in MN/m^2, it is advisable, by a simple mental process, to make that transition at the outset of the solution. Thus, 80 GN/m^2 is seen readily to be 80 000 MN/m^2, which equals 80 000 N/mm^2, and is put in the equation directly in that form.

Although the form MN/m^2 is preferred as the best general form of SI statement, it is of great practical value to note also that MN/m^2 equals N/mm^2.

Ch. 4 STRENGTH OF MATERIALS 97

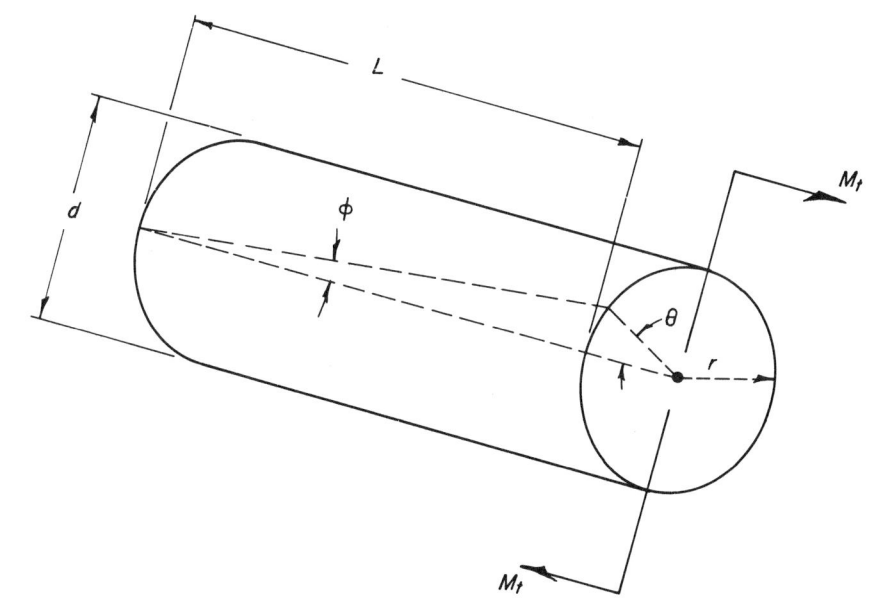

Fig. 4-1. Torsional moment in a solid circular shaft.

Example 4-7

A steel shaft, 1.85 m long, is of hollow circular section, with external and internal diameters of 50 mm and 20 mm respectively. It is subjected to a torsional moment of 130 N·m. If the modulus of rigidity is 80 GN/m², what is the maximum stress and the angle of twist in degrees?

Assuming it is decided to work in millimeters: $M_t = 130\,000$ N·mm. For a hollow shaft:

(Eq. 4.11) $J = (d_0^4 - d_i^4)\pi/32$

$J = (50^4 - 20^4)\pi/32 = 59.8 \times 10^4$ mm⁴

$= 59.8 \times 10^{-8}$ m⁴ $= 598 \times 10^{-9}$ m⁴

Maximum stress occurs at $r = 25$ mm, and is therefore:

$\sigma_s = M_t r/J = 130\,000(25)/59.8 \times 10^4$

$= 5.43$ N/mm² $= 5.43$ MN/m²

$\theta = M_t L/JG = 130\,000(1850)/59.8 \times 10^4 (80\,000)$

$= 0.00503$ rad

$\theta = 0.00503(180/\pi) = 0.288$ degree

4.14 Choice of Unit Multiples

It may be observed that there is little to choose in an example like this between working in meters and in millimeters. Generally the millimeter prefix is treated as a shorthand way of writing 10^{-3}. The expression for the twist in radians might then appear as:

$$\theta = \frac{130(1.85)10^6}{59.8 \times 10^4(80\,000)}$$

$$\frac{\text{N·m(m)}10^6}{\text{mm}^4(\text{N/mm}^2)} = \frac{10^6\,\text{m}^2}{\text{mm}^2} = \frac{\text{mm}^2}{\text{mm}^2} = \frac{\text{mm}}{\text{mm}} = \text{rad}$$

$\theta = 0.005\,03$ rad

In evaluating this expression the units in the numerator need to be changed mentally,

from: (N·m)(m); to (N·mm)(mm) by introducing 10^6,
or, in the denominator

from: mm⁴(N/mm²); to mm²(N); to m²(N), by introducing 10^{-6}.

This procedure needs practice to ensure freedom from mistakes. It is *not* recommended for general use but may be advantageous in certain situations if frequent use assures proper handling.

One advantage of working exclusively in the primary units is that complete consistency is then automatically preserved, so that it becomes unnecessary to write out the units as well. Thus if all the data are first expressed in the primary units they may be written:

$$\theta = M_t L/JG = 130(1.85)/59.8 \times 10^{-8}(80 \times 10^9) = 0.00503 \text{ rad}$$

This procedure gives complete confidence that the answer will be expressed in terms of the primary SI unit for angle, the radian. The same procedure will be used in the following calculation, representative of many arising in the design of shafts.

Example 4-8

A torque of 5000 N·m is transmitted by a coupling in which there are 16 bolts, each 12 mm in diameter, on a pitch circle of a diameter 300 mm. What is the shear stress in the bolts?

Force on each bolt is:	5000/16(0.15) = 2083 N
Shear stress on the bolt:	2083/(0.012)²(0.7854) = 18.4 × 10⁶ N/m²
∴ shear stress is:	18.4 MPa

Ch. 4　　　　　STRENGTH OF MATERIALS　　　　　99

4.15　Beams

The ability of a member to support a load on a given span is usually analyzed by studying separately the transverse shear and bending moment functions. It is customary to present these functions in shear and moment diagrams the shape of which will be unchanged in SI units. Only the notation will change, as shown in the following examples.

In determining the reactions at beam supports it is essential, as previously mentioned under "Statics," to distinguish between a *mass* being carried by the beam and a force applied to the beam.

Example 4-9

A beam carries the loads shown. Neglect the mass of the beam itself. Determine the maximum shear and moment for which the beam must be designed.

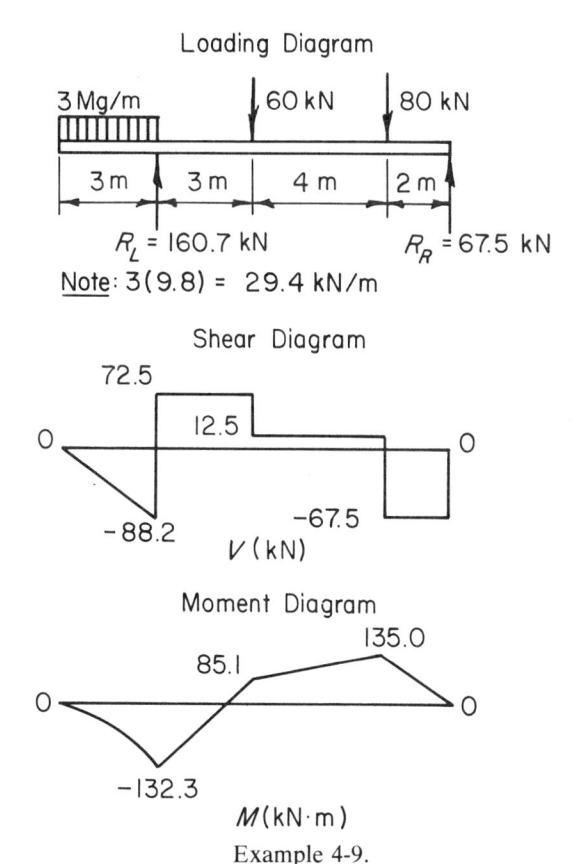

Loading Diagram

3 Mg/m　　　60 kN　　80 kN

3 m　　3 m　　4 m　　2 m

R_L = 160.7 kN　　　R_R = 67.5 kN

Note: 3(9.8) = 29.4 kN/m

Shear Diagram

72.5

12.5

−88.2　　−67.5

V (kN)

Moment Diagram

135.0

85.1

−132.3

M (kN·m)

Example 4-9.

4.16 Fiber Stress in Beams

Fiber stress is determined by the well-known flexure formula, (Eq. 4.12)

$$\sigma = My/I \quad (\text{N·mm})\text{mm}/\text{mm}^4 = \text{N}/\text{mm}^2 = \text{MN}/\text{m}^2 = \text{MPa}$$

or, (Eq. 4.13)

$$\sigma = M/S \quad \text{in which the terms and preferred SI units are:}$$

M = bending moment at section   N·mm

I = second moment of area of cross-section taken about the neutral axis of the beam, which coincides with c.g. axis   mm^4

y = distance to fiber at which stress is being determined; usually termed "c" when it is distance to extreme fiber to obtain maximum unit stress   mm

σ = unit stress in fiber at distance "y" from neutral axis; stress may be either tension or compression depending on sign of bending moment and fiber under consideration   N/mm^2 = MN/m^2 = MPa

S = section modulus; I/c in which "c" is distance to extreme fiber for controlling stress   mm^3

4.17 Second Moment of Area and Section Modulus

The properties of the cross section are, in general, computed from the geometry of the area as indicated in Fig. 4-2.

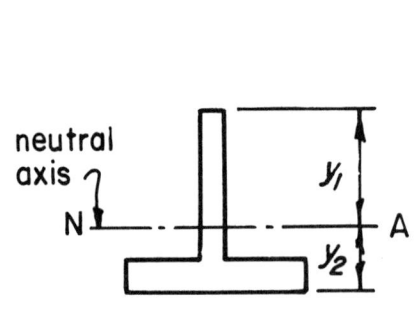

Fig. 4-2. T-Beam properties. 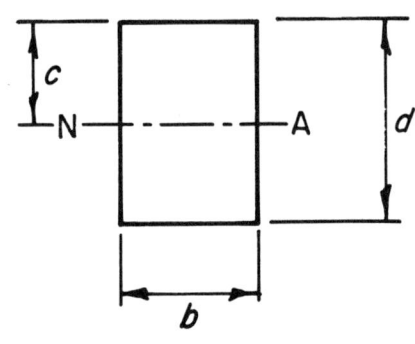   Fig. 4-3. Rectangular beam properties.

Ch. 4 STRENGTH OF MATERIALS 101

In the special case of a rectangular cross section, Fig. 4-3, $I = bd^3/12$ and since $S = I/c$, and $c = d/2$, therefore $S = bd^2/6$.

4.18 Properties of Structural Sections

Values for other special sections, i.e., Structural Steel Shapes, may be obtained from references such as Tables 4-4 and 4-5.

From these tables it should be noted that some structural sections such as wide-flange beams are most likely to experience a "soft" change-over, while other sections, such as structural angles, will experience a "hard" change-over. The reason for this lies in the number of "weights" for each nominal size of wide-flange beam. Since the actual depths may vary considerably from the nominal size there is no point in making a "hard" change-over in sizes. Hence, the properties will merely be published in SI units. There may, of course, at this time of extensive review, be some changes in sizing of the various beam series for the purposes of optimization.

On the other hand there is more reason for "hard" change-over in leg size and thickness of structural angles. This can be more readily accomplished and serve a real purpose since these dimensional considerations are often more specifically a part of a total design.

4.19 SI Units in Beam Analysis

All of the tools for the analysis of fiber stress in beams are now at hand. In its primary form the bending moment is expressed in N·m. However, it is generally preferable to utilize M in terms of N·mm. As has been noted the second moment of the cross-sectional area (I) is mostly likely to be expressed in mm^4 and the section modulus in mm^3.

Although the fourth and third powers involved in the statement of I and S in terms of millimeters may at first seem to result in unwieldy numbers, the judicious use of multipliers in the form of powers of ten does lead to great convenience in calculations. Alternate approaches are shown in the examples which follow.

Example 4-10

A square bar is used as a horizontal cantilever, fixed at one end and subjected to a vertical load of 16 kN at the other. If the bar is 1.2 m long and 90 mm square in section, determine the maximum stress.

One possible procedure is to replace all prefixes by powers of ten and

Table 4-4. Wide-Flange Beams (illustrative) Dimensions and Properties*

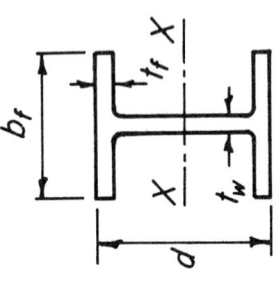

Designation Nominal Size mm	Mass kg/m	Area of Section (A) mm²	Depth of Section (d) mm	Flange Width (b_f) mm	Flange Thickness (t_f) mm	Web Thickness (t_w) mm	About X-X Second Moment (I) 10^6 mm⁴	About X-X Section Modulus (S) 10^3 mm³	About X-X Radius of Gyration (r) mm
W 300 × 46	46.1	5890	307	166	11.8	6.7	99.5	648	130
× 40	40.2	5150	304	165	10.2	6.1	85.2	561	129
W 250 × 37	37.2	4750	256	146	10.9	6.4	55.6	434	108
× 31	31.3	4000	251	146	8.6	6.1	44.4	353	105
W 200 × 30	29.8	3800	207	134	9.6	6.3	28.9	279	87.2
× 25	25.3	3230	203	133	7.8	5.8	23.6	232	85.4

*The values in this table are direct conversions from the customary units of sections presently rolled [soft changeover].

Table 4-5. Unequal Leg Angles (illustrative) Dimensions and Properties*

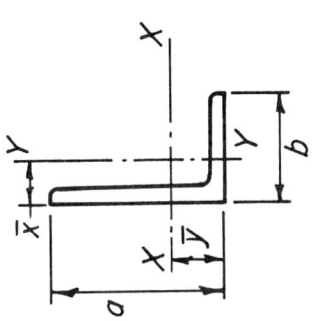

Size and Thickness ($a \times b \times t$) mm	Mass kg/m	Area (A) mm²	Axis X-X (I) 10⁶ mm⁴	Axis X-X (S) 10³ mm³	Axis X-X (r) mm	Axis X-X ($\bar{y}$) mm	Axis Y-Y (I) 10⁶ mm⁴	Axis Y-Y (S) 10³ mm³	Axis Y-Y (r) mm	Axis Y-Y ($\bar{x}$) mm
L 200 × 100 × 15	33.7	4300	17.58	137.0	64.0	71.6	2.99	38.4	26.4	22.2
× 12	27.3	3480	14.40	110.0	64.3	70.3	2.47	31.3	26.7	21.0
× 10	23.0	2920	12.20	93.2	64.6	69.3	2.10	26.3	26.8	20.1
L 100 × 75 × 12	15.4	1970	1.89	28.0	31.0	32.7	0.90	16.5	21.4	20.3
× 10	13.0	1660	1.62	23.8	31.2	31.9	0.78	14.0	21.6	19.5
× 8	10.6	1350	1.33	19.3	31.4	31.0	0.64	11.4	21.8	18.7

*The values in this table are metric equivalents of customary sizes with dimensions adjusted to rationalized values [hard changeover].

to work in terms of the newton and the meter as primary units:

$$M = 16 \times 10^3 (1.2) = 19.2 \times 10^3 \text{ N·m}$$

$$S = \{9 \times 10^{-2}\}^3/6 = 121.5 \times 10^{-6} \text{ m}^3$$

$$\sigma = 19.2 \times 10^3/121.5 \times 10^{-6} = 0.158 \times 10^9 \text{ N/m}^2 = 158 \text{ MPa}$$

Alternatively, the work may be carried out entirely in terms of the newton and the millimeter:

$$M = 16 \times 10^3 (1200) = 19.2 \times 10^6 \text{ N·mm}$$

$$S = (90)^3/6 = 121.5 \times 10^3 \quad \text{mm}^3$$

$$\sigma = 19.2 \times 10^6/121.5 \times 10^3 = 158 \text{ N/mm}^2 = 158 \text{ MPa}$$

The procedure to be followed may be largely one of personal preference. It is however important to be fully aware that the choice does not involve any more or less arithmetic, and that the form in which the answer eventually emerges is usually determined by the preferred general practice. The preferred style is in turn reflected in the form in which basic reference data are given.

Example 4-11

A round bar of mild steel (see Table 4-3) is 200 mm in diameter and 15 m long. Find the maximum stress due to dead load when it is simply supported at the ends in a horizontal position.

Note from the units of mass density (kg/m³) that it will probably be simpler in this instance to work in meters rather than millimeters.

The total mass of the bar is:

$$m = 7850\{(0.2)^2 \pi/4\}\, 15 = 3699 \text{ kg}$$

The gravitational force on the bar is expressed by:

$$F = mg \quad F = 3699(9.8) = 36\,250 \text{ N}$$

For a simple beam with a distributed load:

$$M = FL/8 = 36\,250(15)/8 = 67\,970 \text{ N·m}$$

$$S = \pi d^3/32 = \pi (0.2)^3/32$$

(Eq. 4.14)
$$= 0.000\,785 \text{ m}^3 = 785 \times 10^{-6} \text{ m}^3$$

$$\sigma = M/S = 67\,970/0.000\,785$$

$$= 0.0866 \times 10^9 \text{ N/m}^2 = 87 \text{ MPa}$$

Ch. 4 STRENGTH OF MATERIALS 105

4.20 Beam Deflections

Calculating the deflection of a beam under load involves the modulus of elasticity, which already has been discussed.

Example 4-12

Referring to the bar in the previous example, what will be the deflection at the center of the span?
The deflection formula for this case is:

(Eq. 4.15) $$\delta = \frac{5}{384} \frac{FL^3}{EI}$$

As computed above the gravitational force $F = 36\,250$ N

$L = 15$ m; $E = 200$ GPa   (Table 4-3)

$I = \pi d^4/64 = (0.2)^4 \pi/64$

$\quad = 0.000\,078\,54$ m$^4 = 78.54 \times 10^{-6}$ m^4

$$\delta = \frac{5}{384} \frac{36\,250(15)^3}{200 \times 10^9 (0.000\,078\,54)} \qquad N(m)^3 \cdot \frac{m^2}{N} \cdot \frac{1}{m^4} = m$$

$\delta = 0.101$ m $= 101$ mm

4.21 Handling Units

A wide range of formulae is available for determining second moments, section moduli, bending moments, and deflections. They are all ultimately based on the units of force and length in various combinations, and are self-consistent, in that only one unit is used for each physical quantity: in the English system these are normally the pound (force) and inch. Such formulae may be used unchanged when working in SI units, replacing the pound by the newton and the inch by the meter. Alternatively one may use the newton and millimeter, as already indicated, provided this is done consistently. The last example shows that this procedure should be used discreetly. There is little point, for example, in determining the volume in mm^3 when the density is given in kg/m^3.

The latitude of using either the meter or the millimeter can not be arbitrarily extended to all situations. For instance, in the process of handling calculations relating mass to force, since by definition one

106   STRENGTH OF MATERIALS   Ch. 4

newton (N) is that force which imparts to one kilogram (kg) the acceleration of one meter per second squared (m/s²), the correct numerical value for g is 9.8. If the equivalent linear value of 9800 mm is used then, of course, the resultant force units are millinewtons! Similarly, when acceleration is quantified in terms of m/s², the $\sqrt{g}$ is numerically 3.13; but when it is given in mm/s², the $\sqrt{g}$ is 99.05.

Example 4-13

An aluminum beam of length 12 m with a cross-section in the form of a hollow square with external and internal dimensions 120 mm and 70 mm, respectively, is used as indicated, fixed horizontally at one end and simply supported at the other. Find the maximum stress and maximum deflection due to dead load.

From Table 4-3:

$\rho = 2700 \text{ kg/m}^3 \qquad E = 70 \text{ GN/m}^2 = 70 \text{ GPa}$

$m = \{(0.12)^2 - (0.07)^2\}2700(12) = 307.8 \text{ kg} \qquad\qquad \text{m}^2 \cdot \text{m} \cdot \dfrac{\text{kg}}{\text{m}^3} = \text{kg}$

$F = mg = 307.8(9.8) = 3016 \text{ N} \qquad\qquad \text{kg} \cdot \text{m/s}^2 = \text{N}$

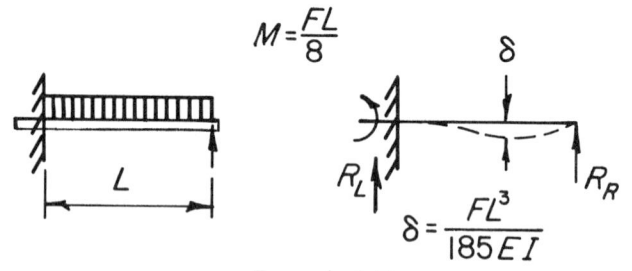

$M = \dfrac{FL}{8}$

$\delta = \dfrac{FL^3}{185EI}$

Example 4-13.

Having found F in newtons we may decide to continue the calculation in terms of the meter, or alternately in terms of the millimeter. Both paths are shown below for the sake of comparison in this example:

Based on $I = bd^3/12$; $c = d/2$; and $S = I/c$

$I = \{120(120)^3 - 70(70)^3\}/12$

$I = 15\,280 \times 10^3 \text{ mm}^4 = 15.28 \times 10^{-6} \text{ m}^4$

$S = 15\,280 \times 10^3/60 = 254.7 \times 10^3 \text{ mm}^3$

$S = 254.7 \times 10^{-6} \text{ m}^3$

Ch. 4 STRENGTH OF MATERIALS 107

Also:
$$E = 70 \times 10^3 \text{ N/mm}^2 = 70 \times 10^9 \text{ N/m}$$

In view of the plan of varying prefixes by intervals of 10^3 it is advisable to keep products in this form as well. Thus the emphasis is on using 10^3, 10^6, etc., and in *not* using 10^2, 10^4, etc. Also note that physical properties such as I and S, can readily be expressed in terms of either m or mm by recording both values with proper exponential notations at the time the initial determination is made. However, there is generally a preference to work with all positive exponents whenever possible.

$$M = FL/8 = 3016(12)/8 = 4524 \text{ N} \cdot \text{m}$$

I—In terms of the meter:

Maximum fiber stress is:

$$\sigma = \frac{M}{S} = \frac{4524}{254.7 \times 10^{-6}} = 17.76 \text{ MPa} \qquad \frac{\text{N} \cdot \text{m}}{\text{m}^3} = \frac{\text{N}}{\text{m}^2} = \text{Pa}$$

Maximum deflection is:

$$\delta = \frac{FL^3}{185EI} = \frac{3016(12)^3}{185(70 \times 10^9)(15.28 \times 10^{-6})} = 0.026 \text{ m} \qquad \frac{\text{N} \cdot \text{m}^3}{(\text{N/m}^2)(\text{m}^4)} = \text{m}$$

II—In terms of the millimeter:

$$\sigma = \frac{M}{S} = \frac{4524 \times 10^3}{254.7 \times 10^3} = 17.76 \text{ N/mm}^2 \qquad \frac{\text{N} \cdot \text{mm}}{\text{mm}^3} = \frac{\text{N}}{\text{mm}^2}$$

but by inspection:

$$\sigma = 17.76 \text{ N/mm}^2 = 17.76 \text{ MN/m}^2 = 17.76 \text{ MPa}$$

$$\delta = \frac{FL^3}{185EI} = \frac{3016(12 \times 10^3)^3}{185(70 \times 10^3)(15\,280 \times 10^3)} = 26 \text{ mm}$$

$$\frac{\text{N} \cdot \text{mm}^3}{(\text{N/mm}^2)\text{mm}^4} = \text{mm}$$

Example 4-14

What uniformly distributed total load may be carried safely by a W 200 × 25 rolled structural steel beam (see Table 4-4), if it is simply supported on a 4.2 m span? What will be the shear stress in the web resulting from this load? Is this safe? (See Table 4-3). What will be the deflection at mid-span? Is this allowable?

108   STRENGTH OF MATERIALS   Ch. 4

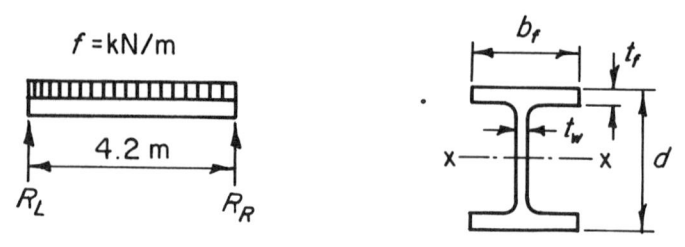

Example 4-14.

From Table 4-4:

$$d = 203 \text{ mm} \qquad t_f = 7.8 \text{ mm}$$
$$t_w = 5.8 \text{ mm} \qquad b = 133 \text{ mm}$$
$$I_{xx} = 23.6 \times 10^6 \text{ mm}^4 \qquad S = 232 \times 10^3 \text{ mm}^3$$

From Table 4-3:

$$\text{allow. } \sigma = 140 \text{ MPa} \qquad \text{allow. } \sigma_s = 50 \text{ MPa} \qquad E = 200 \text{ GPa}$$
$$M = fL^2/8 \qquad \sigma = Mc/I = M/S \qquad \therefore S \cdot \sigma = fL^2/8$$

hence

$$(232 \times 10^{-6})(140 \times 10^6) = f(4.2)^2/8$$
$$(\text{m})^3 \cdot \text{N/m}^2 = (\text{N/m}) \cdot \text{m}^2$$
$$f = 14.73 \text{ kN/m}$$

Checking for shear:

$$V = (14.7)4.2/2 = 30.9 \text{ kN}$$

(Eq. 4.16) $\quad \sigma_s = V/t_w d = 30.9/5.8(203)$

$$= 0.0262 \text{ kN/mm}^2 = 26.2 \text{ N/mm}^2 = 26.2 \text{ MN/m}^2$$

$$\sigma_s = 26.2 \text{ MN/m}^2 = 26.2 \text{ MPa} < 50 \therefore \text{ satis.}$$

Checking for deflections:

$$\delta = \frac{5}{384} \frac{fL^4}{EI} = \frac{5}{384} \frac{14.73 \times 10^3 \, (4.2)^4 \times 10^{12}}{200 \times 10^9 \, (23.6) \, 10^6} = 0.0126 \text{ m}$$

$$\frac{10^3 \text{N}}{\text{m}} \cdot 10^{12} \text{ mm}^4 \cdot \frac{\text{m}^2}{10^9 \text{N}} \cdot \frac{1}{10^6 \text{ mm}^4} = \text{m}$$

Note that when the various quantities are used in the standard formula

Ch. 4   STRENGTH OF MATERIALS   109

in terms of the units and prefixes usually quoted, the answer is the deflection numerically stated in meters. That is:

f ... kN/m, L ... mm, E ... GPa, I ... 10^6 mm^4

yields δ in meters (m), which is readily restated in millimeters, as may be desired.

Actual deflection:

0.0126 m = 12.6 mm;   4200/360 = 11.7 mm allowable

∴ deflection is slightly in excess of usual allowable.

Allowable uniform load in terms of mass:

$m = 14\,730/9.8 = 1503$ kg/m

but beam dead load is

23.5 kg/m ∴ allowable net L.L. is 1480 kg/m

Note in the use of the deflection formula above, the balancing out of exponential values of 10 when the principal components of the equation are expressed as indicated. The load per unit length is in kN/m which is 10^3 N/m. The span is expressed numerically in meters which may then be interpreted as 10^3 mm, giving 10^{12} mm^4 when the span is raised to the fourth power. The modulus of elasticity is used in GPa which is 10^9 N/m^2 and the second moment of the cross section is taken from a reference table giving I values conveniently in 10^6 mm^4. The result is directly a deflection in meters which may, if desired, be readily changed to mm.

4.22 Composite Elastic Elements

When two materials are combined to jointly carry a load, and if corresponding elastic action occurs for each, the unit stresses are proportional to the ratio of the moduli of elasticity.

Example 4-15

A short block of concrete is reinforced with steel bars as shown. What safe compressive load is acceptable?

From Table 4-3:

$E_{st} = 200$ GPa; $E_c = 14$ GPa

elastic ratio

$$n = E_{st}/E_c = 200/14 = 14.3$$

assuming identical elastic action, i.e., $\sigma_{st} = n\sigma_c$ and maximum allowable stresses:

$$\sigma_c = 5.3 \text{ MPa}; \qquad \sigma_{st} = 125 \text{ MPa}$$

but

$$\sigma_{st} = 5.3(14.3) = 75.8 \text{ MPa}$$

when σ_c is maximum allowable.

∴ concrete governs during combined elastic action: 75.8 < 125. Allowing for concrete displaced by steel:

Total allowable load

$$F = A_c\sigma_c + (n - 1)A_{st}\sigma_{st}$$
$$F = 200(300)5.3 + (14.3 - 1)8(201.1)5.3 = 431.4 \text{ kN}$$

As previously mentioned, it is frequently helpful to recognize that $N/mm^2 = MN/m^2$. This results from the fact that:

$$N/mm^2 = N/(10^{-3} \text{ m})^2 \qquad 10^6(N/m^2) = MPa$$

Another point worthy of note is that Table 4-6 has been prepared using the traditional concept of a bar size series based on preferred diameters. An alternate plan is to base the preferred size series on cross-

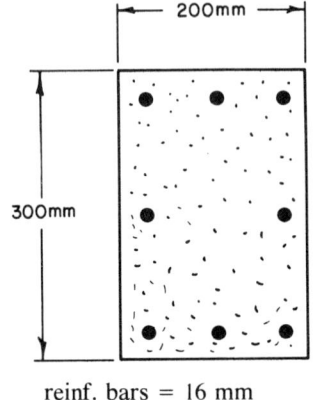

reinf. bars = 16 mm
$A = 201.1$ mm² (each)

Example 4-15.

Table 4-6. Steel Reinforcement for Concrete

Size, mm	Section Area, mm²	Mass, kg/m
6	28.3	0.222
8	50.3	0.395
10	78.5	0.616
12	113.1	0.888
16	201.1	1.579
20	314.2	2.466
25	490.9	3.854
32	804.2	6.313
40	1256.6	9.864
50	1963.5	15.413

Ch. 4 STRENGTH OF MATERIALS 111

sectional areas since the designer is more frequently selecting bars on the basis of area. The latter plan may well emerge as a new standard in SI terms. Table 4-7 gives additional information on size designations, etc.

4.23 Column Action

The various formulas concerned with the buckling of columns are of the same general form regardless of the unit system used, as long as it is consistently applied. For example the Euler formula for the *ultimate* load on a column may be written as:

(Eq. 4.17) $$F = \frac{\pi^2 EI}{CL^2} \qquad \frac{(N/m^2)m^4}{m^2} = N$$

which may alternately be written:

(Eq. 4.18) $$F/A = \frac{\pi^2 E}{C(L/r)^2} \qquad \frac{N/m^2}{(m/m)^2} = N/m^2$$

In this connection it will be remembered that:

(Eq. 4.19) $$r = \sqrt{I/A}$$

In the above formulas symbols have the following meaning and

Table 4-7. Proposed SI Reinforcing Bar Specifications

ASTM Proposed Size Designation	CSA Proposed Size Designation	Nominal Diameter, mm	Nominal Cross-Sectional Area, mm²
1	11	11	100
1.5	—	14	150
2	16	16	200
3	20	20	300
4	—	23	400
5	25	25	500
7	30	30	700
10	36	36	1000
15	44	44	1500
25	56	56	2500

preferred units:

- F = critical buckling load   N
- A = cross-section area m^2
- E = modulus of elasticity N/m^2
- I = second moment of cross-section area m^4
- L = unsupported length   m (or mm)
- r = critical radius of gyration (buckling factor)   m (or mm)
- C = numerical factor depending on end restraint conditions, equal to 1.0 for pin-ended conditions.

Example 4-16

What safe load may be carried by a pair of 100 × 75 × 10 structural angles, placed long legs back-to-back, if used as a pin-ended compression strut with an unsupported length of 3.60 m?

From Table 4-5:

$$a = 100 \text{ mm} \quad b = 75 \text{ mm} \quad t = 10 \text{ mm}$$

(for each angle)

$$\bar{y} = 31.9 \text{ mm} \quad \bar{x} = 19.5 \text{ mm} \quad A = 1660 \text{ mm}^2$$

$$I_{xx} = 1.62 \times 10^6 \text{ mm}^4; \quad I_{yy} = 0.78 \times 10^6 \text{ mm}^4$$

Solving for

$$I_{2-2} = I_{yy} + A(\bar{x})^2$$
$$= 2[(0.78 \times 10^6) + 1660(19.5)^2]$$
$$I_{2-2} = 2.82 \times 10^6 \text{ mm}^4$$

By inspection:

$$I_{xx} = 2(1.62 \times 10^6) = 3.24 \times 10^6 \text{ mm}^4$$

Since $2.82 < 3.24$, $I_{2-2} < I_{xx}$

$$\therefore \text{ least } r = \sqrt{I_{2-2}/A}$$
$$r = (2.82 \times 10^6 / 3320)^{1/2} = 29.1 \text{ mm}$$
$$\max L/r = 3600/29.1 = 123.7$$

∴ Euler's equation is applicable (assuming mild structural steel)

Solving for allowable stress, using Euler's formula with a factor of safety $\phi = 2.5$:

$$\frac{F}{A} = \frac{\pi^2 E}{\phi C(L/r)^2} = \frac{(3.14)^2(200)}{2.5(1.0)(123.7)^2}$$

$$= 0.0516 \text{ GPa} = 51.6 \text{ MPa}$$

∴ total allowable axial load is:

$$51.6 \times 10^6 (2) 1660 \times 10^{-6} = 171.3 \text{ kN}$$

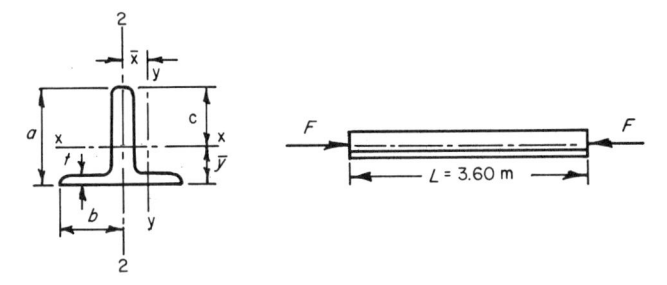

Example 4-16.

Example 4-17

What concentrated load at mid-span may be safely carried by the pair of angles in the diagram if placed with short legs horizontal on a simple span of 2.2 m? See Table 4-5 for "Dimensions and Properties of Structural Angles," and Example 4-16.

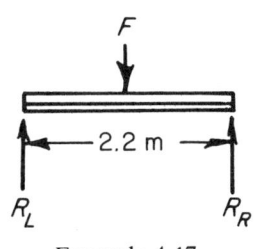

Example 4-17.

From Table 4-3: allow. fiber stress is 140 MPa

$$\sigma = My/I \quad c_y = 100 - 31.9 = 68.1 \text{ mm}$$

$$S = I/y = 2(1.62 \times 10^6/68.1) = 47.6 \times 10^3 \text{ mm}^3$$

$$M = 140(47.6) = 6664 \text{ N} \cdot \text{m}$$

$$= 6.66 \text{ kN} \cdot \text{m} \qquad \frac{10^6 \text{ N}}{\text{m}^2} \cdot 10^3 \text{ mm}^3 = \text{N} \cdot \text{m}$$

$$M = FL/4 \quad F(2.2/4) = 6.66 \quad F = 12.12 \text{ kN}$$

4.24 Direct Stress with Bending

The following example introduces two additional principles but does not involve any novelty from the point of view of units.

Example 4-18

The diagram shows a clamp frame of rectangular section. It is required to determine the maximum stresses at the section A-B resulting from the clamp force of 4 kN

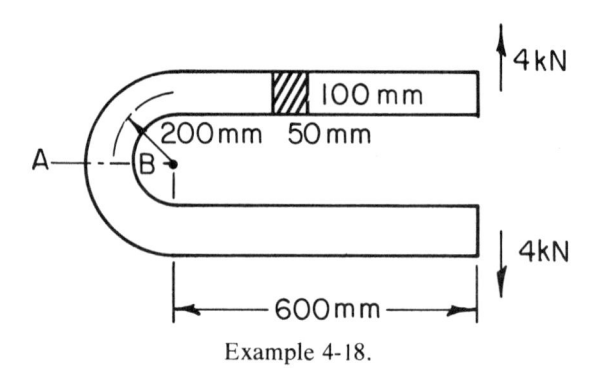

Example 4-18.

The bending moment at AB is

$$4000(600 + 200) = 3\,200\,000 \text{ N} \cdot \text{mm} = 3.2 \times 10^6 \text{ N} \cdot \text{mm};$$

the distance of the mid-point of the section from A or B is 50 mm; the second moment of the cross-section area is:

$$I = 50(100)^3/12 = 4.17 \times 10^6 \text{ mm}^4$$

Hence, using the ordinary formula for a straight beam, the stresses at

A and B would be:

$$\sigma = Mc/I = 3.2 \times 10^6(50)/4.17 \times 10^6 = 38.4 \text{ N/mm}^2 = 38.4 \text{ MPa}$$

The true stresses at A and B are, however, modified by the curvature and a resulting shift of the neutral axis. Formulae available in the handbooks indicate that the straight-beam stresses should be multiplied by factors of 0.85 for A and 1.20 for B. These are pure numbers, determined solely by ratios of linear dimensions, which would normally be expressed in millimeters. The true stresses due to bending A and B are, accordingly, 38.4(0.85) = 32.6 MPa (in compression) and 38.4(1.20) = 46.1 MPa (in tension), respectively.

To those must be added, algebraically, the direct stress at AB due to the clamping force; this will be a tensile stress of:

$$\sigma = F/A = 4000/50(100) = 0.8 \text{ N/mm}^2 = 0.8 \text{ MPa}$$

The final stress at A therefore, is: 32.6 − 0.8 = 31.8 MPa in compression, and at B is: 46.1 + 0.8 = 46.9 MPa, in tension.

4.25 Lateral Strain

Consideration of the lateral distortion of elastic elements introduces Poisson's ratio, which is defined as lateral strain divided by longitudinal strain; it is, therefore, of neutral dimensions.

Example 4-19

A metal bar of rectangular cross-section 40 mm by 20 mm is subjected to a tension of 600 kN. If Young's modulus is 110 GPa and Poisson's ratio is 0.34, what will be the resulting reduction in cross-sectional area of the bar?

The longitudinal unit strain will be:

$$\epsilon = \frac{F}{AE} = \frac{600\,000}{20 \times 40(110\,000)} = 0.0068 \text{ mm/mm} \qquad \frac{N}{mm^2} \cdot \frac{mm^2}{N} = \frac{mm}{mm}$$

The lateral strain will therefore be 0.0068(0.34) = 0.0023 mm/mm and the reduction in cross-sectional area will be (20)40(0.0023)² = 0.0042 mm².

Example 4-20

A metal cube of edge 250 mm is immersed in the sea to a depth producing a pressure of 3.0 MPa.

If Young's modulus for this metal is 205 GPa and Poisson's ratio ν is 0.29, what will be the reduction in volume?

This involves the bulk modulus, K, which is defined as stress/volumetric strain; volumetric unit strain is of neutral dimensions, so that the primary unit for K is the same as that for stress; namely, the newton per square meter. It can be shown that for these conditions:

$$K = \frac{E}{3(1 - 2\nu)} = \frac{205}{3(1 - 0.58)} = 163 \text{ GPa} = 163\,000 \text{ MPa}$$

The volumetric strain resulting from a pressure of 3.0 MPa is therefore equal to:

$$\sigma/K = 3.0/163\,000 = 18.4 \times 10^{-6} \text{ mm}^3/\text{mm}^3 \qquad \text{MPa/MPa}$$

so that the change in volume is:

$$(250)^3(18.4 \times 10^{-6}) = 287 \text{ mm}^3.$$

4.26 Other Dimensional Effects

The formulae and methods used for calculating stresses and dimensional changes in shells, both thin- and thick-walled, involve various combinations of linear dimensions, forces, stresses, elastic moduli, and Poisson's ratio, but do not introduce any new units. The following example will illustrate the effect of centrifugal force in which the dynamic effects of rotation are introduced.

Example 4-21

A hollow cylinder of 150-mm diameter and 2-mm wall thickness contains air at atmospheric pressure; it is rotating about its axis in a vacuum at a speed of 9000 r/min. Find the hoop stress in the wall if the density of the material is 7500 kg/m³.

It can be shown that the hoop stress due to rotation is $\rho \bar{V}^2$, and that due to internal pressure is pr/t, where ρ is the density, $\bar{V}$ the linear velocity, p the internal pressure, r and t, radius and wall thickness, respectively. Further $\bar{V} = r\omega$, where ω is the angular velocity. Common formulae of this type can be used without modification with the primary SI units; note in particular that the angular velocity is in rad/s.

Here the pressure p is the excess of "atmospheric" over "vacuum," and in the absence of more precise information it will be sufficient to assume that the external p is approximately: -100 kPa. The total hoop stress is accordingly:

$$\sigma_t = \rho \bar{V}^2 + pr/t = \rho(r\omega)^2 + pr/t$$

$$\sigma_t = 7500[(0.075)(2\pi)(9000/60)]^2$$
$$+ 100 \times 10^3(0.075/0.002)$$

$$\frac{kg}{m^3}\left[\frac{m}{rad} \cdot \frac{rad}{r} \cdot \frac{r}{min} \cdot \frac{min}{s}\right]^2 + \frac{N}{m^2} \cdot \frac{mm}{mm} = \frac{N}{m^2} + \frac{N}{m^2} = \frac{N}{m^2} = Pa$$

$$\sigma_t = (37.47 + 3.75)10^6 = 41.2 \text{ MPa}$$

Note the pedantry involved here in evaluating r/t; It is obviously unnecessary to change the units since they cancel at once. Note also that the presence of the radian signals involvement with a dynamic condition.

4.27 Springs

The following formulas are typical of those encountered in the design of springs. For a close-coiled helical spring of circular wire, the deflection is:

(Eq. 4.20) $$\delta = \frac{8FND^3}{Gd^4} \qquad \frac{N \cdot m^3}{(N/m^2)m^4} = m$$

and the maximum shear stress is:

(Eq. 4.21) $$\sigma_s = \frac{8FD}{\pi d^3} \qquad \frac{N \cdot m}{m^3} = \frac{N}{m^2}$$

Here the symbols have the following significance: F is the load, in newtons; N is the number of active coils; D is the diameter of the coil, in meters; G is the modulus of rigidity, in pascals; d is the diameter of the wire, in meters; the deflection is given in meters and the shear stress in pascals.

From the above it will be seen that, as usual, it is possible also to use the millimeter consistently instead of the meter.

Example 4-22

A helical compression spring is made of round wire 2 mm in diameter; the material has a modulus of rigidity of $G = 70$ GPa. There are 20 working coils, of 18 mm mean diameter. What will be the deflection and maximum stress when the coil is called upon to support a weight of 0.3 kg, and what will the strain energy be in the spring?

The gravitational force is 0.3(9.8) = 2.94 N. Observing that 70 GPa = 70 000 MN/m² = 70 000 N/mm², and working in millimeters, the deflection will be:

$$\delta = \frac{8(2.94)20(18)^3}{70\,000(2)^4} = 2.45 \text{ mm} \qquad N \cdot mm^3 \cdot \frac{mm^2}{N} \cdot \frac{1}{mm^4} = mm$$

The shear stress is:

$$\sigma_s = \frac{8(2.94)18}{\pi(2)^3} = 16.8 \text{ N/mm}^2 = 16.8 \text{ MPa}$$

The "strain energy" in such a spring is equal to the work done in producing the distortion. Here a force of average value 2.94/2 = 1.47 N moves a distance of 2.45 mm, so that the work done, and therefore the strain energy, equals:

$$(2.94/2)0.00245 = 0.0358 \text{ N} \cdot \text{m} = 0.036 \text{ J}$$

A force which is suddenly applied produces a stress greater than that resulting from the steady application of the same force. For example, a general formula for determining the shock stress produced by dropping a mass on a spring is:

$$\text{shock stress} = \text{static stress} \{1 + (1 + 2h/y)^{1/2}\}$$

where h is the distance the mass falls and y the deflection produced by the same weight applied as a static load. Evidently, if h and y are expressed in the same unit, e.g., both in millimeters, this kind of expression produces no problem

Formulae of a more specific type may introduce the customary "weight" itself, and here it is necessary to observe that it is a force and *not* a mass which is intended.

Example 4-23

In Example 4-22, if the "mass" had been dropped on the spring from a height of 0.025 m, what would have been the maximum stress?

The static stress was found to be 16.8 MPa, and the static deflection 2.45 mm. The "mass" falls 25 mm and the shock stress is therefore:

$$16.8\{1 + (1 + 50/2.45)^{1/2}\} = 16.8(5.63) = 94.5 \text{ MPa}$$

5

Mechanics of Machines

5.1 Power Ratings

Machines are intimately concerned with the transmission of power and terms such as "power rating" come at once to mind. As has already been explained, the sole SI unit for power is the watt (W); it is exactly the same as the joule per second (J/s) or the newton-meter per second (N·m/s) which in terms of the base SI units is the kg·m²/s³. One of the major achievements of the International System is that in this all-important matter of the utilization of energy, all branches of technology are brought together in the use of a common familiar unit—the watt—for power. See Table 5-1, "Quantities and SI Units Used in Mechanics of Machines," and Table 5-2, "Formulas Frequently Used in Mechanics of Machines."

5.2 The Joule, The Ampere, and The Watt

The key to this vital, straightforward relationship lies in the identification of the ampere (A) as a Base Unit of SI and its definition (see Fig. 8-1) in terms of force units, newton (N), as follows:

> The *ampere* is that constant current which, if maintained in two straight parallel conductors of infinite length, of negligible cross section, and placed one meter apart in vacuum, would produce between these conductors a force equal to 2 × 10⁻⁷ newton per meter of length. (CIPM-1946, Resolution 2, approved by the 9th CGPM-1948, Resolution 7.)

This definition of the ampere thus provides a direct linkage between electrical energy and mechanical energy and accordingly establishes a common basis for all energy and power "ratings." Making this observation entirely from a "units" point of view it may be stated that:

$$N·m = J \qquad J/s = W \qquad J = W·s \qquad W = A·V$$

Table 5-1. Quantities and SI Units Used in Mechanics of Machines

Quantity Symbol	Quantity Name	SI Unit Name	SI Unit** Symbols
α	angular acceleration		rad/s^2
D, d	diameter, or depth		m; mm
δ	deflection		mm
E	modulus of elasticity	pascal	Pa
E_k	kinetic energy	joule	J
F	force	newton	N
g	acceleration of gravity		m/s^2
G	modulus of rigidity (shear modulus of elasticity)	pascal	Pa
I	second moment of cross-sectional area		m^4; mm^4
$\bar{I}$	inertia, rotational		kg·m^2/rad^2
J	polar second moment of cross-sectional area		m^4; mm^4
L	length	meter	m; mm
m	mass	kilogram	kg
n	speed, rotary	revolutions per second	r/s
n	frequency (cycles per second)	hertz	Hz
P	power	watt	W
ρ	mass density		kg/m^3
r	radius		m; mm
$\mathfrak{r}$	gyrational factor		m/rad
σ_t	tensile stress	pascal	Pa
σ_s	shear stress	pascal	Pa
M_t	torsional moment (static)	newton-meter	N·m; N·mm
S_p	polar modulus		m^3; mm^3
T	torque (kinetic)	newton-meter per radian	N·m/rad
θ	angle; angle of twist		rad; rad/m
γ	angle		rad
μ	coefficient of friction		N/N
$\bar{V}$	velocity, linear		m/s
ω	velocity, angular*		rad/s

* Usually observed or quoted in terms of "revolutions," i.e.: r/min; r/s.
** For complete interpretation of this table, note the following in terms of Base Units:

$N = kg \cdot m/s^2$ $\qquad$ $J = N \cdot m = kg \cdot m^2/s^2$ $\qquad$ $W = J/s$

$Pa = N/m^2 = kg/(m \cdot s^2)$ $\qquad$ $Hz = 1/s$ $\qquad$ $W = kg \cdot m^2/s^3$

Ch. 5   MECHANICS OF MACHINES   121

Table 5-2. Formulas Frequently Used in Mechanics of Machines

5.1	$P = T\omega$	5.8	$n = (\pi/2L^2)(EI/m)^{1/2}$
5.2	$P = F\bar{V}$	5.9	$n = (1/2\pi)\sqrt{GJ/IL}$
5.3	$P = 2\pi nT$	5.10	$\omega_c = (kEI/mL^3)^{1/2}$
5.4	$D = (8P/\pi^2 n \sigma_s)^{1/3}$	5.11	$\delta = mgL^3/kEI$
5.5	$E_k = \frac{1}{2} m \underline{r}^2 \omega^2 = \frac{1}{2} m \bar{V}^2$	5.12	$\omega_c = (g/\delta)^{1/2}$
5.6	$\sigma_t = \rho \underline{r}^2 \omega^2$	5.13	$P = U_p \bar{V} f d$
5.7	$\Omega = M_t / I \omega$	5.13a	$P = U_p(\bar{V}fd)/60$

Illustrations of power associated with linear motion have already been given in Chapter 3, Dynamics. Machines, however, are more often concerned with rotary motion as is illustrated in the Examples which will follow.

Example 5-1

A shaft rotating at a speed of 50 rad/s is transmitting a torque of 120 N·m/rad. What is the power transmitted?

(Eq. 5.1) $P = T\omega = 120(50) = 6000$ W $\dfrac{\text{N·m}}{\text{rad}} \cdot \dfrac{\text{rad}}{\text{s}} = \text{N·m/s} = \text{W}$

$P = 6.0$ kW

It is worthwhile contrasting this simplicity in computation with comparable calculations in customary English units. If the speed of rotation were given in r/min and the torque in ft-lbf, it would be necessary to multiply the product by $2\pi/33\,000$ to obtain the answer in horsepower, which then could be converted into watts by multiplying by 746. The need for such conversion factors is, for the most part, eliminated in SI.

5.3 Angular Measure in Rotating Machinery

In rotating machinery, in view of the direct relationship between each revolution of the rotating part and events of a cyclical nature, the customary approach is to cite speed of rotation in revolutions per unit of time. Because of the frequent need for information about fractions of a cycle, it is common practice to use the degree as the customary subdivision of a single revolution. Thus, from an applied point of view, revolution (r) and degree (°) will continue to be the principal units of data in observations concerning rotary motion. But in design computations, especially those involving energy and power as explained in Chapter 3, Dynamics, the radian must be utilized.

Thus, SI is consistent with the design requirement by identifying the radian as the key unit of angular measure. This also means that the factor 2π radians/revolution will continue to appear in many formulas applied to rotary motion. It is possible, of course, by a simple modification in dial read-out, to have a tachometer which will record rotational speed in radians per unit of time, but since numerical conversion in design computations is so readily done, it seems unlikely that such devices will come into common field usage.

Wide use of the second (s) as the reference unit of time for rotary motion (r/s) will save much computation.

Example 5-2

A belt passes without slipping over a pulley of diameter 750 mm; the linear speed of the belt is 25 m/s. If the tensions on the tight and slack sides are respectively, 110 N and 80 N, what power is being transmitted?

The power transmitted is the product of the net force and the velocity:

(Eq. 5.2) $\quad P = F\bar{V} = (110 - 80)25 = 750 \text{ W} = 0.75 \text{ kW} \quad\quad \text{N·m/s} = \text{W}$

Note that the diameter of the pulley is immaterial when the linear speed of the belt is given.

Example 5-3

If the arc of contact of the belt with the pulley is 200° what will be the minimum coefficient of friction to avoid slippage?

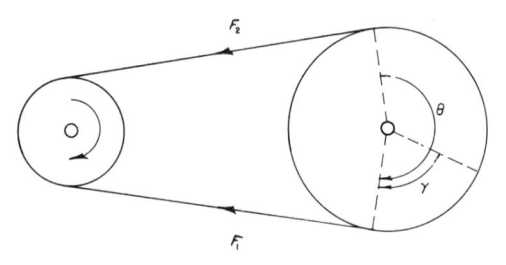

Example 5-3.

Ch. 5   MECHANICS OF MACHINES   123

The usual expression for the above relationship is: $F_1/F_2 = e^{\mu\gamma}$ in which γ is the arc over which creep is occurring. Obviously, when γ equals θ slippage occurs. Therefore the minimum required coefficient of friction μ to avoid slippage is:

$$\mu = \frac{1}{\theta} \log_e \frac{F_1}{F_2}$$

$$= \frac{1}{200(\pi/180)} \log_e \left(\frac{110}{80}\right) \quad \mu = 0.091$$

Note that for consistent units θ must be expressed in radians.

Example 5-4

If the speed of rotation of the pulley in the previous examples were increased by 100 r/min, other things remaining equal, what would be the increase in the power transmitted?

Net torque is:

$$T = \Delta F \underline{r} = 30(0.75/2) = 11.25 \text{ N·m/rad}$$

Change in speed:

$$\omega = 100(2\pi/60) = 10.47 \text{ rad/s}$$

Increase in power

$$P = 11.25(10.47) = 117.8 \text{ W} \qquad \frac{\text{N·m}}{\text{rad}} \cdot \frac{\text{rad}}{\text{s}} = \frac{\text{J}}{\text{s}} = \text{W}$$

5.4 Power Shafting

Some calculations are so commonplace that, despite the simplicity of SI units, it is still worthwhile to use a "pre-digested" formula, particularly when the physical properties of conventional materials are involved. Such formulas do involve a new set of numerical factors when expressed in SI and some new considerations in respect to handling of SI units may become involved. To illustrate, consider the development of an equation (for torsion only) which will permit direct solution for the required diameter of a power-transmitting shaft given:

P = power to be transmitted W

n = rotational speed of shaft r/s

D = diameter of solid circular shaft m

T = torque (kinetic) N·m/rad

M_t = torsional moment (static) N·m

Equations (4.8) and (4.10) may be used to develop the necessary *static* relationships:

$$M_t = \sigma_s(J/r) \qquad J/r = S_p \qquad S_p = \pi D^3/16$$

In a parallel consideration the equation for power transmitted by a rotating machine may be used to express the dynamic terms as:

(Eq. 5.3) $$P = 2\pi n T$$

These two separate analyses are best viewed first in the following parallel comparison:

Statically:

$$\sigma_s = \frac{M_t}{S_p} \qquad \frac{\text{N·m}}{\text{m}^3} = \frac{\text{N}}{\text{m}^2}$$

$$M_t = \sigma_s \cdot S_p \qquad \frac{\text{N}}{\text{m}^2} \cdot \text{m}^3 = \text{N·m}$$

Dynamically:

$$P = 2\pi n T \qquad \frac{\text{rad}}{\text{r}} \cdot \frac{\text{r}}{\text{s}} \cdot \frac{\text{N·m}}{\text{rad}} = \frac{\text{N·m}}{\text{s}}$$

$$T = \frac{P}{2\pi n} \qquad \frac{\text{N·m}}{\text{s}} \cdot \frac{\text{r}}{\text{rad}} \cdot \frac{\text{s}}{\text{r}} = \frac{\text{N·m}}{\text{rad}}$$

A proper interpretation of the above is that the static stress σ_s is the result of the physical dimensions of the shaft and the torsional moment M_t only. The static stress occurs whether or not there is rotation. Since the magnitude of the static stress is *not* dependent on rotation it remains constant during any period of rotation at constant speed. The interfacing with the equation for P, is simply a device for determining the numerical value of M_t. Thus, it should be expected that the unit check-out may yield a residual radian unit which is merely a reflection of the procedure used. It should not enter the actual description of the final answer of the static solution which is, therefore, simply in terms of the meter.

Stated in another way, the dynamic torque T may be evaluated as N·m on an instantaneous basis and thereby, for the purpose of computing static stress, may be equated numerically to static torsional moment M_t, which is normally given in N·m. This is the same as observing that what is being sought, by equating static to dynamic conditions, is simply the magnitude of the torsional moment M_t which, *if* it were maintained throughout the period of rotation, would give rise to the quantity of power (P) indicated.

Accepting these conditions it is possible to proceed as follows:

Since $\quad \sigma_s S_p = P/2\pi n, \quad$ then: $\quad \sigma_s \cdot \dfrac{\pi D^3}{16} = \dfrac{P}{2\pi n}$

This relationship (for torsion only) can be expressed in the general form:

(Eq. 5.4) $$D = \sqrt[3]{\dfrac{8P}{\pi^2 n \sigma_s}}$$

Example 5-5

From a torsional point of view only, what would be a suitable diameter of a power shaft to transmit 150 kW at 500 rpm?

Assume
$$\sigma_s = 27.5 \text{ MPa}$$

Note that
$$n = 500/60 = 8.33 \text{ r/s}$$

Using Equation (5.4):

$$D = \left[\dfrac{8(150 \times 10^3)}{\pi^2(8.33)27.5 \times 10^6} \right]^{1/3}$$

$$D = 0.0809 \text{ m} \quad D = 80.9 \text{ mm}$$

For convenience such formulas sometimes are further simplified by inserting an assumed allowable stress such as $\sigma_s = 27.5$ MPa and recognizing that $S_p = 0.196 D^3$.

The result is:

$$D^3 = \frac{P}{27.5\ (2\pi)n(0.196)} = \frac{29\,500\ P}{n}$$

$$D = \sqrt[3]{\frac{29\,500\ P}{n}}$$

In the above equation the numerical value of 29 500 reflects the insertion of the magnitude of σ in terms of the multiple MPa, a numerical multiple factor of 10^6, and the anticipation of the insertion of the magnitude of P in terms of the multiple kW. Hence the result will be D directly in millimeters!

Such specialized formulas should be used only after a thorough analysis of the assumptions made and of the units or unit multiples presumed. For instance, in the above equation the unit check-out would be as follows:

Noting that $\qquad$ W = J/s = N·m/s

the units for D^3 are:

$$\text{kW} \cdot \frac{1}{\text{MPa}} \cdot \frac{\text{rev}}{\text{rad}} \cdot \frac{\text{s}}{\text{rev}} = \frac{\text{N·mm}}{\text{s}} \cdot \frac{\text{mm}^2}{\text{N}} \cdot \text{s} = \text{mm}^3$$

The presence of the radian is a reminder of the fact that this result has been achieved by equating a static condition to a dynamic condition and is true only as long as this equivalence applies during all increments of rotation under consideration.

For further discussion of the "Compatible Unit Multiples" see Section 4.13. Also see the following discussion on a specialized formula for both bending and torsion in shafts where it is indicated that a consistent set of unit multiples may be used in one way or another.

5.5 Bending and Torsion in Shafts

It is well recognized that the actual design of shafts generally involves bending as well as torsion. Combined bending and torsion may be indirectly designed for by using a substantially reduced shear stress in the torsional analysis as was done above.

Preferably, the combined conditions of bending and torsion would be handled by a formula such as the following:

$$d^3 = \frac{16\pi}{\sigma_s[(k_m M)^2 + (k_t M_t)^2]^{1/2}}$$

in which:

	Alternate Unit Sets	
	(1) or,	(2)
D, d = shaft diameter	m	mm
σ_s = design shear stress ($0.3\sigma_{yt}$ or, $0.18\sigma_{vt}$)	Pa = N/m²	MPa = MN/m² = N/mm²
M = bending moment	N·m	N·mm
M_t = torsional moment (static)	N·m	N·mm
k_m = shock factor (1.5 for gradually applied load)	...	...
k_t = fatigue factor (1.0 for gradually applied load)	...	...

In other words, such formulas may be used directly if applied with a consistent set of units, either (1) or (2), and dimension-free numerical constants.

Torsional Deflection is often a controlling factor in the final selection of a shaft size. Such problems involve a determination of the angle of twist θ, which was discussed in Section 4.12 and Example 4-7. Deflections due to bending were similarly discussed in Sections 4.20 and Example 4-12.

5.6 Gears

The design of gears is very much a matter of linear dimensions and strength-of-materials calculations which have already been covered in an earlier chapter.

In describing procedures for checking fine-toothed gears it is usual to specify the "pressures" to be used, but these quantities are more correctly termed "forces," and thus should be expressed in newtons. The power transmitted is then found from the product of the torque and the rate of rotation.

128 MECHANICS OF MACHINES Ch. 5

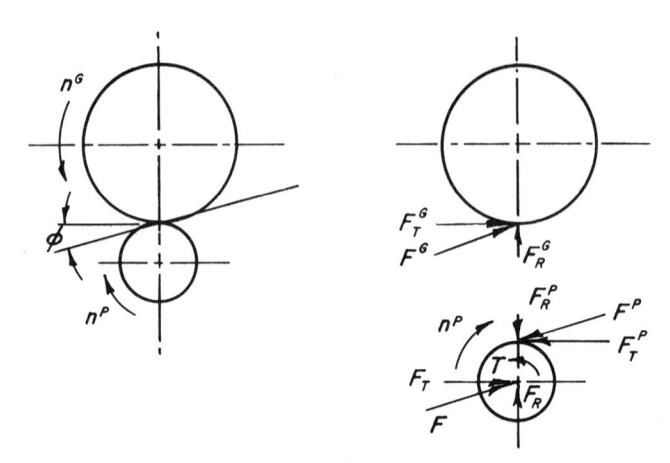

Example 5-6.

Example 5-6

A 20-tooth, 20-degree full-depth pinion is rotating at 1725 r/min and transmitting 4.0 kW. If the "module" for this gear is 3.5 (millimeters of pitch diameter to number of teeth), calculate the force F^P on the teeth of the pinion, the bearing load on the bearing, and the torque to be transmitted by the pinion shaft.

1. The pitch diameter is: $D_p = 3.5(20) = 70$ mm
2. The pitch line velocity is: $\pi(70)1725/60 = 6322$ mm/s
3. The tangential tooth load F_T^P is obtained by observing that:

$$F_T^P(6.322) = 4000 \qquad \text{N·m/s} = \text{W}$$

$$F_T^P = 632.7 \text{ N}$$

4. The resultant tooth load is:

$$F^P = \frac{F_T^P}{\cos 20°} \qquad F^P = 673.3 \text{ N}$$

As can be seen from the sketch, the resultant tooth load is also the bearing load.

5. The torque transmitted by the pinion shaft is:

$$T = F_T^P(D_p/2) = 632.7(35)$$

$$= 22\,145 \text{ N·mm/rad} = 22.145 \text{ N·m/rad}$$

Ch. 5   MECHANICS OF MACHINES   129

Example 5-7

Assume that the above torsional moment is applied through a 30-mm steel shaft, what is the stress in the shaft and the angle of twist in a length of 500 mm? Assume, $G = 80$ GPa $= 80\,000$ MN/m^2 $= 80\,000$ N/mm^2. Using Equations (4.8) and (4.10):

$$\sigma_s = M_t \div J/r = 22\,145 \div \pi(30)^3/16$$

$$\sigma_s = 4.177 \text{ N/mm}^2 = 4.18 \text{ MN/m}^2 = 4.18 \text{ MPa}$$

Applying Equation (4.9):

$$\theta = M_t L/JG$$

$$\theta = \frac{22\,145(500)}{80\,000\{\pi(30)^4/32\}} = 0.00174 \text{ rad}$$

$$(\text{N·mm})(\text{mm}) \cdot \frac{\text{mm}^2}{\text{N}} \cdot \frac{1}{\text{mm}^4} = \frac{\text{mm}}{\text{mm}} = \text{rad}$$

Example 5-8

A spur pinion rotates at 400 rad/s; the pitch diameter is 60 mm and the tangential load is 50 N. What is the power transmitted?

$$P = 50(0.03)400 = 600 \text{ W} = 0.6 \text{ kW} \qquad \frac{\text{N·m}}{\text{rad}} \cdot \frac{\text{rad}}{\text{s}} = \frac{\text{J}}{\text{s}} = \text{W}$$

Example 5-9

An electric motor running at 1800 r/min is to drive a centrifugal pump at 600 r/min through a pair of spiral bevel gears. The required (kinetic) torque to be exerted by the motor will be 30 N·m/rad. What power does this represent?

$$P = 2\pi(1800/60)30 = 5655 \text{ W} = 5.66 \text{ kW}$$

$$\frac{\text{r}}{\text{min}} \cdot \frac{\text{rad}}{\text{r}} \cdot \frac{\text{min}}{\text{s}} \cdot \frac{\text{N·m}}{\text{rad}} = \text{N·m/s} = \text{J/s} = \text{W}$$

A successful design must ensure that the stresses in the teeth are not excessive, so that they neither break nor wear at an unacceptable rate. For conventional types of gear, formulae may be found in handbooks linking the many parameters involved. Despite their apparent complexity these formulae are composed of nothing more than lengths, rates of rotation, and stresses together with certain numerical factors which can be obtained from charts. In the formulae devised for working in SI units, an essential difference is that the numerical factors for converting between foot-pounds per second and horsepower are no longer involved.

5.7 Flywheels

The purpose of most flywheels is to equalize the energy demand by maintaining, at a more or less constant level, the total of the energy applied and the energy stored. Thus, a flywheel prevents excessive changes of speed in processes which involve the intermittent application of considerable forces. Calculations are largely concerned with the amount of energy which corresponds to a given change in the speed of the flywheel. This elementary calculation has already been illustrated in Chapter 3, Dynamics, particularly in Example 3-14. The energy associated with a flywheel having a moment of inertia of $\bar{I}$ (expressed in kg·m²/rad²) rotating at an angular velocity ω(rad/s), is equal to $½\bar{I}\omega^2$, expressed in joules.

Many flywheels consist essentially of a heavy rim supported by comparatively light spokes or by a thin web. As a close approximation, the principal rotational energy effective in flywheel action *in such cases* is considered to be:

(Eq. 5.5) $\qquad E_k = ½m\bar{r}^2\omega^2, \quad$ or, $\quad E_k = ½m\bar{V}^2$

in which it is taken that $\bar{V} = \bar{r}\omega$, and that:

$\quad m$ = mass of the rim $\qquad$ kg
$\quad \bar{r}$ = gyrational factor (mean radius of rim) $\qquad$ m/rad
$\quad \omega$ = rotational speed (angular velocity) $\qquad$ rad/s
$\quad \bar{V}$ = linear velocity of rim $\qquad$ m/s

Example 5-10

A shearing machine makes 20 strokes per minute; each cut takes 0.75 second and involves a force of 1.2 MN acting through a distance of 110 mm. Design a flywheel having a mean diameter of 2.5 m, running at 200 r/min, such that the reduction in speed is not more than 10 percent per stroke.

Energy absorbed per stroke:

$\qquad 1.2 \times 10^6 (110 \times 10^{-3}) = 132$ kJ $\qquad$ N·m = J

Ch. 5   MECHANICS OF MACHINES   131

Total time interval per stroke:

$$60/20 = 3 \text{ s}$$

Power to be generated for each stroke:

$$132/3 = 44 \text{ kW} \qquad \text{J/s} = \text{W}$$

During the intervals between strokes this power is being used solely to accelerate the flywheel, so that the energy buildup in the flywheel is:

$$44(3 - 0.75) = 99 \text{ kJ} \qquad \text{W} \cdot \text{s} = \text{J}$$

This must equal the energy lost by the flywheel in slowing down during the cut, which may be written as:

$$E_k = \tfrac{1}{2}mr^2(\omega_1^2 - \omega_2^2)$$
$$= \tfrac{1}{2}m(1.25)(\{200\}^2 - \{180\}^2)(2\pi/60)^2$$

$$E_k = 65.1m \text{ joules} \qquad \text{kg} \cdot \frac{\text{m}^2}{\text{rad}^2} \cdot \frac{\text{rad}^2}{\text{s}^2} = \text{kg} \cdot \frac{\text{m}^2}{\text{s}^2} = \text{N} \cdot \text{m} = \text{J}$$

thus:

$$m = 99\,000/65.1 = 1521 \text{ kg} = 1.521 \text{ Mg} = 1.52 \text{ metric ton.}$$

Since flywheels are frequently operated at high speeds, an important factor in their design is the ability to withstand the centrifugal stresses produced. Example 3-16 indicates the type of calculation involved, but another example may be helpful.

Example 5-11

A flywheel consists essentially of a rim of cross section 80 mm × 20 mm with a mean diameter of 800 mm, and rotates at 1000 r/min. Find the maximum hoop stress in the material if its density is 8 Mg/m³.

Treating the rim as "thin," the formula for the hoop stress σ_t does not involve its cross section provided the density is given.

(Eq. 5.6) $\sigma_t = \rho r^2 \omega^2$

$$\sigma_t = 8 \times 10^3 (0.4)^2 (1000\{2\pi/60\})^2$$

$$= 14\,040\,000 \text{ N/m}^2 \qquad \frac{\text{kg}}{\text{m}^3} \cdot \frac{\text{m}^2}{\text{rad}^2} \cdot \frac{\text{rad}^2}{\text{s}^2} = \frac{\text{kg}}{\text{m}} \cdot \frac{1}{\text{s}^2} = \frac{\text{N}}{\text{m}^2}$$

$$\sigma_t = 14.04 \text{ MN/m}^2 = 14.04 \text{ MPa}$$

132   MECHANICS OF MACHINES   Ch. 5

5.8 Gyroscopes

Calculations concerning gyroscopes include some of the same aspects of unit considerations as in the flywheel examples.

Example 5-12

A simple gyroscope includes a flywheel having a rotational inertia $\bar{I}$ of 1.8 kg·m²/rad² and rotating at 400 rad/s with a horizontal axis of spin. A couple M_t of 220 N·m is applied to the framework in a direction which will cause the axis to tilt. Find the rate of precession of the system.

Utilizing the applicable mathematical formula, the rate of precession Ω is found as follows:

(Eq. 5.7) $\Omega = \dfrac{M_t}{\bar{I}\omega} = \dfrac{220}{1.8(400)} = 0.3$ rad/s $\dfrac{\text{kg·m}^2}{\text{s}^2} \cdot \dfrac{\text{rad}^2}{\text{kg·m}^2} \cdot \dfrac{\text{s}}{\text{rad}} = \dfrac{\text{rad}}{\text{s}}$

5.9 Vibrations

Vibration is essentially concerned with simple harmonic motion, and this has already been briefly illustrated in Example 3-27, which dealt with the oscillation of a mass suspended from a spring. Many other types of oscillation arise in machinery but, as will be demonstrated in the following examples, the calculations involved do not introduce any novel factors as far as units are concerned.

Example 5-13

A simply supported horizontal solid steel bar of uniform rectangular section has a length of 5 m, a depth of 150 mm, a width of 100 mm, and a mass of 90 kg/m. What is the fundamental frequency for free oscillations in the vertical plane?

(Eq. 5.8)   The frequency is given by: $n = (\pi/2L^2)\sqrt{EI/m}$

L = span   m
m = mass per unit of length   kg/m
E = stretch modulus of elasticity   Pa
I = second moment of cross-sectional area   m⁴

For steel, $E = 200$ GPa;

$$I = bd^3/12 = 0.10(0.15)^3/12 = 28.1 \times 10^{-6} \text{ m}^4$$

$$n = \{\pi/2(5)^2\}\sqrt{200 \times 10^9 (28.1 \times 10^{-6})/90} \qquad n = 15.7 \text{ Hz}$$

$$\frac{1}{m^2}\left(kg \cdot \frac{m}{s^2} \cdot \frac{1}{m^2} \cdot m^4 \cdot \frac{m}{kg}\right)^{1/2} = \frac{1}{s} = \text{Hz}$$

Example 5-14

A flat disc weighing 18 kg has a diameter of 260 mm. It is suspended centrally from a steel wire 2 mm in diameter and 1.2 m long, and is set to oscillate in a horizontal plane. Find the frequency of the oscillations.

The frequency in this case is given by:

(Eq. 5.9) $\qquad n = (\tfrac{1}{2}\pi)\sqrt{GJ/\bar{I}L}$

in which:

G = modulus of rigidity of the wire   Pa
J = polar second moment of the cross-sectional area of the wire   m^4
L = length of wire   m
$\bar{I}$ = inertia of the disc (rotational)   kg·m²/rad²

$$J = \pi d^4/32 = \pi(0.002)^4/32 = 1.57 \times 10^{-12} \text{ m}^4 \text{ (wire)}$$
$$\bar{I} = mr^2/2 = 18(0.26)^2/8 = 0.152 \text{ kg·m}^2/\text{rad}^2 \text{ (disc)}$$

for steel wire: $G = 85$ GPa, whence

$$n = \tfrac{1}{2}\pi (85 \times 10^9\{1.57 \times 10^{-12}\}/0.152\{1.2\})^{1/2} = 0.136 \text{ Hz}$$

$$\frac{r}{rad}\left(kg \cdot \frac{m}{s^2} \cdot \frac{1}{m^2} \cdot m^4 \cdot \frac{rad^2}{kg \cdot m^2} \cdot \frac{1}{m}\right)^{1/2} = \frac{r}{rad} \cdot \frac{rad}{s} = \frac{r}{s} = \frac{cycle}{s} = \frac{1}{s} = \text{Hz}$$

In alternate terms this vibration is $0.136(60) = 8.2$ oscillations per minute.

5.10 Limiting Torque

Calculations concerning the operation of clutches involve considerations such as the acceleration of rotating bodies and torques having limiting values.

Example 5-15

A machine is driven by a motor through a clutch which can exert a maximum torque of 55 N·m/rad. The rotating parts of the machine have a total moment of inertia of 4.5 kg·m/rad². If the clutch is engaged when the machine is at rest, what will its speed be after 8 seconds, assuming that the clutch is still functioning at its maximum torque? Applying Equation (3.9):

$$T = \bar{I}\alpha$$

$$\alpha = T/\bar{I} = 55/4.5$$

$$= 12.2 \text{ rad/s}^2 \qquad \left(kg \cdot \frac{m}{s^2} \cdot \frac{m}{rad}\right)\left(\frac{rad^2}{kg \cdot m^2}\right) = rad/s^2$$

After 8 seconds the angular velocity will be:

$$12.2(8) = 97.8 \text{ rad/s} \quad \text{or,} \quad 97.8(60/2\pi) = 934 \text{ r/min}$$

5.11 Whirling Shafts

Calculations concerning the whirling of shafts involve centrifugal forces and the modulus of elasticity of the material. It may be shown that the critical or whirling speed ω_c of a shaft of length L, bearing a disc of mass m, is given by:

(Eq. 5.10) $$\omega_c = \sqrt{\frac{kEI}{mL^3}}$$

where E and I respectively are the stretch modulus of the material and the second moment of the cross-sectional area of the shaft, and k is a numerical factor which depends upon the position of the disc and the way the bearings are mounted.

It will be recognized that k is the same factor as that used in the customary formula for δ, the static deflection of a shaft due to a concentrated load. Also E, I, and L are the same quantities. It is only the appearance of m and g to represent the load on the shaft that is different in the SI form of the equation. This leads to:

(Eq. 5.11) $$\delta = \frac{mgL^3}{kEI}$$

By substitution it may be written that:

(Eq. 5.12) $$\omega_c = \sqrt{g/\delta}$$

To further pursue this point it will be instructive to examine the units, since these formulae provide another insight into the distinction between mass and weight

The whirling of the shaft results from centrifugal forces which do not involve gravity. But the static deflection results from a gravitational force which does involve g. Hence, when the substitution of the whole expression for deflection is made to obtain the simplified formula for the critical whirling speed ω_c, gravity must in effect be neutralized (or extracted) by introducing an off-setting g.

In detail, the proper units in each case are:

$$\omega_c = \sqrt{\frac{kEI}{mL^3}} \qquad \left(\frac{kg \cdot m}{s^2} \cdot \frac{1}{m^2} \cdot m^4 \cdot \frac{1}{kg} \cdot \frac{1}{m^3}\right)^{1/2} = \frac{m}{m} \cdot \frac{1}{s} = \text{rad/s}$$

$$\omega_c = \sqrt{g/\delta} \qquad \left(\frac{m}{s^2} \cdot \frac{1}{m}\right)^{1/2} = \left(\frac{m}{m}\right)^{1/2} \cdot \frac{1}{s} = \frac{m}{m} \cdot \frac{1}{s} = \text{rad/s}$$

As previously discussed it is essential that the presence of the radian be noted wherever rotary motion is involved.

Example 5-16

A shaft of 500-mm length and 15-mm diameter is mounted on freely supported bearings at each end. It carries a disc of mass 20 kg, fixed at 200 mm from one end. Find the critical speed.

Using the static deflection approach:

$$\delta = \frac{Wa^2b^2}{3EIL}$$

in which W is the gravitational force ($m \cdot g$) in newtons (N); a and b are the respective distances of the disc from the ends of the shaft, i.e.: 0.2 and 0.3, respectively.

$$L = 0.5 \text{ m}$$

Assume:

$$E = 210 \text{ GPa}$$
$$I = \pi D^4/64 = \pi(0.015)^4/64 = 2.48 \times 10^{-9} \text{ m}^4$$
$$\delta = \frac{20(9.8)(0.2)^2(0.3)^2}{3(210 \times 10^9)(2.48 \times 10^{-9})0.5}$$
$$= 0.904 \times 10^{-3} \text{ m} = 0.904 \text{ mm}$$
$$\omega_c = \sqrt{g/\delta} = (9.8/0.904 \times 10^{-3})^{1/2}$$
$$= (10\,851)^{1/2} = 104.2 \text{ rad/s}$$

Note that to convert $\sqrt{g/\delta}$ from rad/s to r/s, a further modification of the formula is sometimes made by introducing 2π, and since $\sqrt{g} = \sqrt{9.8}$ is 3.13, which is almost equal to π, the expression is occasionally written as: $\omega_c = \frac{1}{2}\sqrt{1/\delta}$ in r/s. *Obviously considerable discretion and caution is needed in utilizing such short cuts*, since the unit check-outs may become confused and the real meaning of the phenomenon being studied is detrimentally obscured.

5.12 Governors

The operation of a centrifugal governor involves a somewhat similar interaction between mass and weight; mass is involved in the centrifugal forces which tend to lift the rotating masses, whereas weight is involved in the gravitational forces tending to restore them to the rest position. The forces resulting from the extension of springs are also frequently involved in governor calculations. Illustrations of how to handle the units in all of these situations have already been dealt with in Dynamics.

5.13 Machining Power Requirements

There are many instances in which essential data now expressed in customary units will need to be converted to equivalent SI units. An example is the data on power requirements for machining. The equation below is for turning, planing, and shaping; by making a small change it may also be used for milling or any other machinery operation except grinding:

(Eq. 5.13) $$P = U_p \cdot \bar{V} f d$$

where:

	Alternate Unit Sets	
	(1), or,	(2)
P = power at the cutter	W	kW
$\bar{V}$ = cutting speed	m/s	(m/min)*
f = feed rate (per cycle)**	mm	mm
d = depth of cut	mm	mm
U_p = unit power factor	W·s/mm³	kW·s/1000 mm³

Values of U_p in SI terms may be obtained by multiplying the customary data (usually given in "unit horsepower" which has been defined as "horsepower per cubic inch per minute") by the factor of 2.73 obtained in the following manner:

$$\left(\text{hp}\cdot\frac{\text{min}}{\text{in.}^3}\right)\frac{1}{16\,387}\cdot\frac{\text{in.}^3}{\text{mm}^3}\,(60)\,\frac{\text{s}}{\text{min}}\,(746)\,\frac{\text{W}}{\text{hp}} = 2.73\text{ W}\cdot\text{s/mm}^3$$

which alternately is:

$$2.73 \text{ kW}\cdot\text{s}/1000 \text{ mm}^3$$

A few representative values of U_p are:

AISI 1020 steel (150–175 *Bhn*)	1.58 W·s/mm³, or, kW·s/1000 mm³
AISI 4140 steel (175–200 *Bhn*)	1.36 W·s/mm³, or, kW·s/1000 mm³
Gray Cast Iron (150–175 *Bhn*)	0.82 W·s/mm³, or, kW·s/1000 mm³
Hard Bronze	2.27 W·s/mm³, or, kW·s/1000 mm³

Based on the coherence of SI and its decimal characteristics, these same numerical factors may be applied in terms of kW·s/cm³.

Example 5-17

Determine the power required at the cutter to turn a part made of AISI 2335 steel with a Brinnel hardness number of 220, if the cutting

---

* Use Equation (5.13a).
** This is essentially the width of thickness of cut which when multiplied by the "depth" and the "cutting speed" will give a volume of material removed in one minute. In turning, this would be the feed rate per revolution of the workpiece to maintain the same thickness of cut; in planing and shaping it would be the feed rate per stroke; in general it is the feed rate per cycle, whether continuous or intermittent.

138   MECHANICS OF MACHINES   Ch. 5

speed is to be 70 m/min, the depth of cut 2.54 mm, and the feed rate 0.20 millimeters per revolution.

Solution (1):

From reference tables $U_p = 1.69$ W·s/mm^3, and the turning operation planned in this case calls for application of a feed correction factor $C = 1.08$. Utilizing a modified form of Equation (5.13):

$$P = U_p C \bar{V} f d = 1.69(1.08)(70\,000 \div 60)(0.20)2.54$$
$$= 1082 \text{ W} = 1.082 \text{ kW}$$

Since this is the required "power at the cutter," the efficiency of the machine itself must, of course, be considered in determining the total power requirement.

Solution (2):

In some practical applications it may be desired to adjust the above approach to accommodate data as used or given for particular situations. Thus the alternate formula is:

(Eq. 5.13a) $P = U_p C (fd\bar{V}/60)$ Using Unit Set (2)

$$P = 1.69(1.08)\frac{0.20(70)2.54}{60} = 1.082 \text{ kW}$$

$$\frac{\text{kW·s}}{1000 \text{ mm}^3} \cdot \text{mm} \cdot \text{mm} \cdot \frac{1000 \text{ mm}}{\text{min}} \cdot \frac{\text{min}}{\text{s}} = \text{kW}$$

5.14 Counterbalancing

Calculations in SI for determining the suitable mass and location for counterbalancing are demonstrated by the following:

Example 5-18

The dynamic effect of two cranks positioned 90 degrees apart is 500 kg for M_1 and 600 kg for M_2. Counterbalance weights, M_A and M_B, in planes A and B, respectively, are to be positioned at a radius of 500 mm, the radius of the weights M_1 and M_2 is 250 mm. Calculate the magnitudes and angular positions of M_A and M_B.

MECHANICS OF MACHINES

Example 5-18.

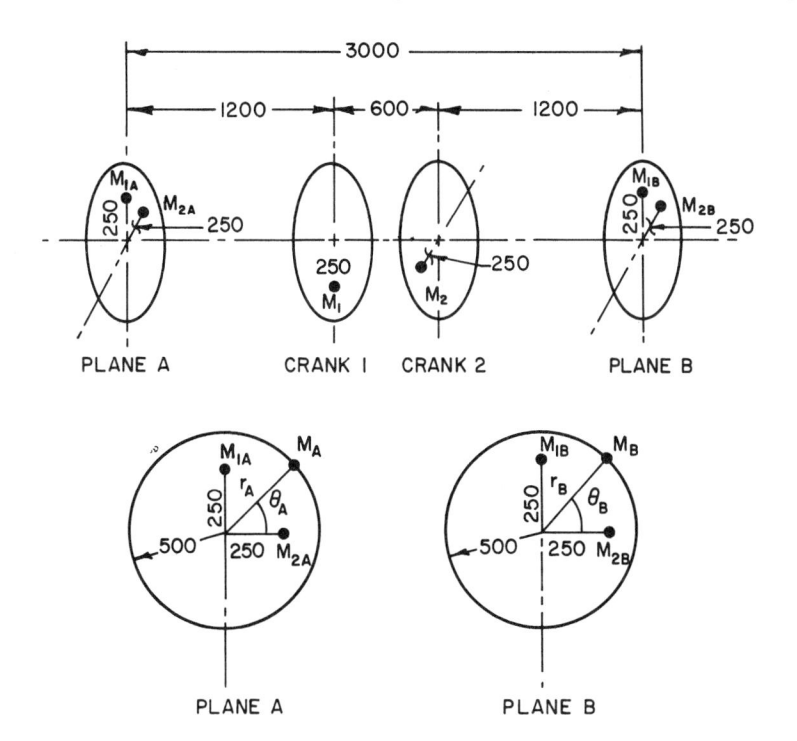

Note – All dimensions in millimeters (mm).

In order to determine M_A and M_B in planes A and B, respectively, moment equations are written around planes A and B.

Σ Moments $= \Sigma mr\omega^2 L$. However, since ω^2 is constant, this equation becomes: $\Sigma mrL = 0$

if the counterbalance masses are to balance the dynamic effect of the cranks.

Although several other methods of solution are possible, one of the best is to consider each weight (M_1 and M_2) separately and to calculate a separate balance weight for each (M_{1A}, M_{1B}, M_{2A}, M_{2B}) in the planes A and B. A further simplification is now possible by considering the radius of the counterbalance weights to be equal to the radius of the crank weights, which reduces the basic equation to a simple moment equation: $\Sigma mL = 0$. The effect is shown in the separate views of planes A and B.

140   MECHANICS OF MACHINES   Ch. 5

Now to solve the problem:

Σ moments @ A = 0 $M_{1B} = 500(1200/3000) = 200$ kg

$M_{2B} = 600(1800/3000) = 360$ kg

Σ moments @ B = 0 $M_{1A} = 500(1800/3000) = 300$ kg

$M_{2A} = 600(1200/3000) = 240$ kg

Each plane is now considered separately as a problem of balancing in a single plane using $\Sigma\, mr\omega^2$.

Thus in plane A, resolving force effects into x and y components:

$$M_A \cos \theta_A = 240(250/500) = 120$$

$$M_A \sin \theta_A = 300(250/500) = 150$$

The resultant mass and angle of position is then:

$$\sqrt{(M_A \cos \theta_A)^2 + (M_A \sin \theta_A)^2} = \{120^2 + 150^2\}^{1/2}$$

$$M_A = 192.1 \text{ kg}$$

$$\tan \theta_A = \frac{M_A \sin \theta_A}{M_A \cos \theta_A} = \frac{150}{120} = 1.25$$

$$\theta_A = 51.34°$$

in a similar manner, the result of B is found to be:

$M_B = 205.9$ kg $\tan \theta_B = 0.556$ $\theta_B = 29.05°$

6

Fluid Mechanics

6.0 Scope of Fluid Mechanics

Many engineering calculations are concerned with relatively incompressible liquids at constant temperature. These conditions usually constitute the major considerations under the more general heading of "Fluid Mechanics" and will be treated in this chapter. The compressible flow of gases, which may be accompanied by substantial temperature changes, will be treated chiefly in Chapter 7, "Thermodynamics and Heat Transfer."

Table 6-1 and Table 6-2, respectively, give "Quantities and SI Units" and the "Formulas" most frequently used in the area of Fluid Mechanics, as treated in this chapter.

The two particular quantities chiefly involved in fluid mechanics are density and pressure, which have already been described, together with viscosity and surface tension. There is a proper SI expression for each of these derived from the Base Units in accordance with the applicable physical dimensions. The discussion and examples which follow will be used to establish these relationships.

6.1 Density, Specific Weight, and Specific Gravity

Before entering the actual discussion of fluid mechanics it will be helpful to review the relationships between three physical characteristics: density, specific weight, and specific gravity.

Density, ρ, is mass per unit volume. In SI, density is therefore expressed in kg/m^3. For pure water, for instance, the standard density value at (4°C) is conveniently 1000 kg/m^3. For general engineering purposes, the density of water does not vary appreciably. The density of several other substances is given in appropriate reference tables as well as in Table 6-3.

Table 6-1. Quantities and SI Units Frequently Used in Fluid Mechanics

Quantity Symbol	Quantity Name	SI Unit Name	SI Unit* Symbols
A	area	square meter	m^2; mm^2
d	diameter	meter	m; mm
δ	deflection	millimeter	mm
e_m	energy per unit of mass	...	J/kg
E	total energy	joule	J
E_p	potential energy, pressure	joule	J
e_p	potential energy, pressure per unit volume	...	J/m^3
E_z	potential energy, elevation	joule	J
e_z	potential energy, elevation per unit volume	...	J/m^3
f	friction factor (Darcy)	...	...
f'	friction factor (Fanning)	...	...
F	force (incl. pressure force)	newton	N
g	acceleration due to gravity	...	m/s^2
h	head	meter	m; mm
L	length	meter	m
m	mass	kilogram	kg
$\dot{m}$	mass rate of flow	...	kg/s
P	power	watt	W
p	unit pressure	pascal	Pa
ρ	mass density	...	kg/m^3
σ	surface tension	...	N/m
Q	volumetric flow rate	...	m^3/s; L/s
R	Reynold's Number	...	neutral dimensions
R	hydraulic radius	meter	m
s	specific gravity	...	...
t	time	seconds	s
$\bar{V}$	velocity	...	m/s
V	volume	...	m^3
ν	kinematic viscosity	...	m^2/s
μ	dynamic viscosity	pascal-second	Pa·s
w	specific weight	...	N/m^3
z	elevation above datum	meter	m

* For complete interpretation of this table, note the following in terms of Base Units:

$N = kg \cdot m/s^2$ $J = N \cdot m = kg \cdot m^2/s^2$ $W = J/s$

$Pa = N/m^2 = kg/(m \cdot s^2)$ $Pa \cdot s = N \cdot s/m^2 = kg/(m \cdot s)$ $W = kg \cdot m^2/s^3$

Table 6-2. Formulas Frequently Used in Fluid Mechanics

6.1	$p = \rho g h$	6.16	$\dfrac{\bar{V}_1^2}{2g} + z_1 + h_1 - h_{f1-2} + \dfrac{\Delta e_m}{g}$ $= \dfrac{\bar{V}_2^2}{2g} + z_2 + h_2$
6.2	$F = pA$		
6.3	$F_b = V\rho g$		
6.4a	$E_p = pv$		
6.4b	$E_p = pm/\rho$	6.17	$P = \dot{m}gh$
6.4c	$E_p = mgh$	6.18a	$P = E/t = mgh/t$
6.5	$e_p = \rho g h$	6.18b	$P = pm/\rho t = p\dot{m}/\rho$
6.6	$E_z = mgz$	6.19	$Q = \dfrac{c_d A_2 \sqrt{2g\Delta h}}{\sqrt{1 - (A_2/A_1)^2}}$, or, $Q = K\sqrt{\Delta h}$
6.7	$e_z = \rho g z$		
6.8	$e_p + e_z = \rho g(h + z)$		
6.9	$E_p + E_z = mg(h + z)$		
6.10	$\delta = 4\sigma/\rho g d$		
6.11	$Q = A\bar{V}$		
6.12	$\dot{m} = Q\rho$	6.20	$F = \dot{m}\Delta\bar{V} = Q\rho\,\Delta\bar{V}$
6.13	$\dfrac{m\bar{V}^2}{2} + mgz + \dfrac{mp}{\rho}$ = constant	6.21	$h_{f1-2} = \dfrac{32\,\nu\bar{V}L}{gd^2} = \dfrac{p_1 - p_2}{\rho g}$
6.14	$\dfrac{\bar{V}^2}{2} + gz + \dfrac{p}{\rho}$ = constant	6.22	$R = \dfrac{\bar{V}d\rho}{\mu} = \dfrac{\bar{V}d}{\nu}$
6.15	$\dfrac{\bar{V}_1^2}{2g} + z + h$ = constant	6.23	$f = 64/R$
		6.24a	$p_1 - p_2 = 2f'\rho\bar{V}^2 L/d$
		6.24b	$p_1 - p_2 = \tfrac{1}{2}f\rho\bar{V}^2 L/d$
		6.25a	$h_{f1-2} = f\dfrac{L}{d}\dfrac{\bar{V}^2}{2g}$
		6.25b	$f = 0.316/R^{0.25}$
		6.26	$P = Q\rho g h$

Specific weight, w, is weight per unit volume, which is the gravitational force effect of a unit volume. This is expressed as N/m³ (or kN/m³) in SI terms. For water at normal temperatures $w = 9810$ N/m³ $= 9.81$ kN/m³. This is equivalent in U.S. Customary Units to 62.4 lbf/ft³. In general the relationship is: $w = \rho g$.

Specific gravity, s, is the ratio at the density of any substance to that of pure water at standard temperature, generally 15°C. Note that the concept of specific gravity yields an interesting value in SI computations since the mass density of water in SI is 1000. Thus the density in SI terms, when expressed as Mg/m³ is synonymous with "specific gravity" in customary terms; for instance, the density of mercury is 13.57 Mg/m³. The preferred statement of density, however, is in kg/m³ which is 1000

Table 6-3. Physical Properties of Some Fluids
(at 15°C and standard atmospheric pressure, illustrative)

Fluid	Density, ρ kg/m³	Specific Weight, w kN/m³	Specific Gravity, s	Dynamic Viscosity, μ mPa·s	Kinematic Viscosity, $\nu = \mu/\rho$ mm²/s	Surface Tension* (in air), σ mN/m	Vapor Pressure, p_v kN/m²
Air	1.23	0.01205	0.00123	0.0178	14.5	...	...
Fresh Water	1000	9.81	1.00	1.14	1.14	73	1.72
Sea Water	1030	10.10	1.03	1.20	1.17	74	1.72
Crude Oil	860	8.44	0.86	16.0	18.6	28	39.3
Gasoline	730	7.16	0.73	0.562	0.77	25	68.9
Mercury	13 570	133.1	13.57†	1.627	0.12	481	0.00010

* Data on surface tension may be encountered expressed in terms of the former metric units, dyne/cm. Since the dyne is 10^{-5} N, the dyne/cm = 10^{-3} N/m. This is often written as μ N/mm.

† Some sources use s = 13.5951

times the customary statement of specific gravity. That is, the mass density of mercury in SI is expressed as 13 570 kg/m³, see Table 6-3. Hence, in SI the term "specific gravity" becomes superfluous.

6.2 Pressure

The concept of "pressure" has its roots in the subject area of hydrostatics which deals with the conditions of fluids at rest.

Consider in a fluid with a "free" surface, an internal vertical column of fluid which extends to a depth h where there is an imaginary horizontal surface of area A (see Fig. 6-1). Since the unit pressure p at any point in a fluid is the same in all directions, and is normal to any immersed surface, the pressure on the sides of the "column" will produce only horizontal forces in equilibrium. Hence the surface bounding the lower end must provide a vertical force of magnitude sufficient to balance the gravitational effect on the mass of the entire fluid column. Accordingly:

ρ = mass density .... kg/m³
h = height of fluid .... m
g = acceleration due to gravity .... m/s²
p = pressure per unit area .... N/m²

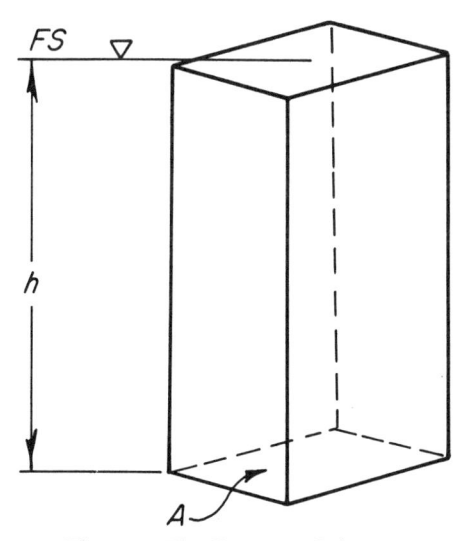

Fig. 6-1. The Pressure Column.

whence $$p \cdot A = \rho g (h \cdot A)$$

(Eq. 6.1) $\quad p = \rho g h \quad\quad \dfrac{\text{kg}}{\text{m}^3} \cdot \dfrac{\text{m}}{\text{s}^2} \cdot \text{m} = \text{kg} \cdot \dfrac{\text{m}}{\text{s}^2} \cdot \dfrac{1}{\text{m}^2} = \dfrac{\text{N}}{\text{m}^2} = \text{Pa}$

The special name "pascal" (symbol Pa) has been given to the SI Derived Unit for pressure. It will be helpful to remember that one pound force per square inch (1.00 psi) is equivalent to: 6.895 kPa (approx. 7 kPa).

At this point it might be well to observe that pressures generally, unless otherwise specified, are taken to be "gage" pressures, i.e., pressures in respect to atmospheric pressure.

When "absolute pressures" are to be used, as in thermodynamic computations, pressures stated must include atmospheric pressure. In SI, standard atmospheric pressure (one atmosphere) equals 101.3 kPa. In some instances it will be desired to state a measurement of "vacuum"; such a condition should be stated as a negative pressure differential in respect to the atmosphere.

As will be noted from Fig. 6-2, gage pressures and vacuums are measured in respect to atmospheric pressure which is a variable datum. Atmospheric pressure changes with elevation on the earth's surface and with meteorological conditions. Absolute pressure, on the other hand, is measured in respect to a fixed datum.

Traditionally, for selective reasons, pressure measurements have been stated in a variety of terms notably: pounds per square inch, pounds per square foot, inches of mercury, feet of water, inches of water, bars, millibars, etc. The ISO recommendation is that all pressures be stated in terms of the "pascal" (newton per square meter) using a choice of one

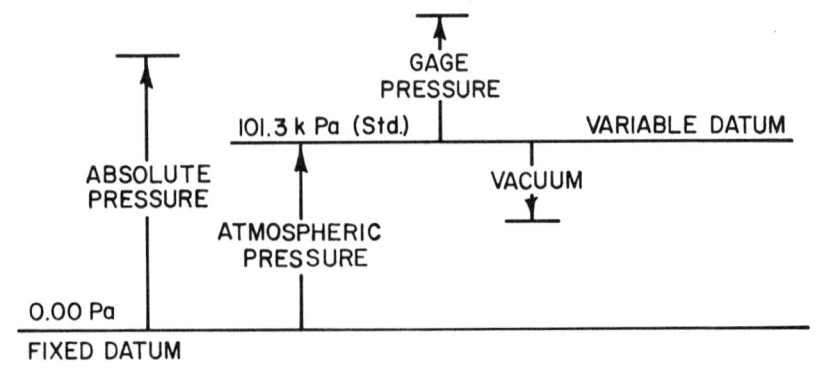

Fig. 6-2. Pressure concepts and terminology.

of the following prefixes as may be appropriate:

> low pressure ranges: Pa, mPa, μPa
> intermediate pressures: kPa
> high pressures: MPa, GPa

Preferred usage, as recommended by ANMC, generally favors the kPa.

Of course, pressure measuring devices can continue to be used which will read "mm of mercury" and "mm of water" or "m of water," but it probably will be found desirable to convert most of these (or the data from them) at an appropriate time, to a multiple of Pa. The conversion factors for doing this are:

> *1.0 mm mercury = 0.133 08 kPa = 133.08 Pa;
> *1.0 mm water   = 0.009 807 kPa =   9.807 Pa;
> *1.0 m water   = 9.807   kPa = 9807   Pa;
>   1.0 lbf/in.^{2} = 6.8948   kPa = 6894.8   Pa.

Note how readily the shift can be made from kPa to Pa, or other multiples of Pa, without the introduction of any other numerical factors. The same kind of shift can readily be made when readings are taken in meters instead of millimeters. The great advantage of the versatility of SI is again self-evident.

When it is necessary to distinguish definitely between gage and absolute pressure, the use of parenthetical identification is recommended. Thus:

	Quantity Symbol	Unit Symbol
Gage Pressure:	p(gage)	Pa
Absolute Pressure:	p(abs)	Pa

Note that the place to identify a pressure as "gage" or "absolute" is with the Quantity Symbol and *not* with the Unit Symbol, since the units are the same whether the pressure is gage or absolute. Of course, a pressure can be stated in text material either as: p(abs) = 27 kPa; or as a pressure of 27 kPa (abs). But the notation in parentheses is a part of the text and *not* a part of the unit symbol.

In order to avoid any possible confusion with the pascal symbol, the use of subscripts to distinguish absolute pressure from gage pressure is disapproved.

The most suitable prefix is one which will give the most frequent values between 0.1 and 999 for the operating range. For further

---

* These factors will vary with temperature and gravity.

information on "pressure," see Section 1.6.8 and also "Preferred Usage in SI and Conversions," in Chapter 9.

Example 6-1

What is the liquid pressure 170 m below the surface in sea water?

$\rho = 1030$ kg/m³ $p = \rho g h$

$p = 1030(9.8)170 = 1\,716\,000$ N/m² $= 1.716$ MPa (gage)

This is approximately 17 "atmospheres." In terms of absolute pressure this would be:

$$1.716 + 0.1013 = 1.817 \text{ MPa (abs)}$$

Example 6-2

A room is closed by a door of height 2.2 m and width 0.8 m. If the air pressure inside the room is increased by 500 Pa (about 2″ of water) what will be the resultant force on the door?

Note that in a fluid of low density, such as air, at or near atmospheric pressure, the variation of pressure within the given height is negligible. Also note that neither gravity nor density appears in this calculation since the pressure is stated directly. Thus:

F = total force resulting from fluid pressure   N
p = pressure per unit area   N/m² = Pa
A = area   m²

(Eq. 6.2) $F = pA = 2.2(0.8)500 = 880$ N $= 0.88$ kN $\text{m} \cdot \text{m} \cdot \dfrac{\text{N}}{\text{m}^2} = \text{N}$

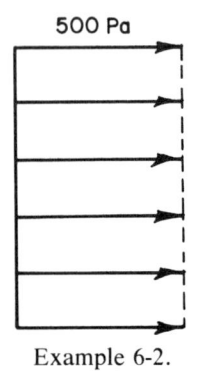

500 Pa

Example 6-2.

Ch. 6   FLUID MECHANICS   149

Example 6-3

Assume that the room and door in the previous example are replaced by a tank and a closure of the same dimensions. If the tank is filled just to the top of the closure with fresh water, what will be the resultant force on the closure? As previously established, unit pressure varies directly with depth, hence at any depth y, the total force will be:

$$F = \int \rho g y(x)\, dy$$

$$= \int_0^{2.2} 1000(9.8)0.8 y\, dy = 19\,000 \text{ N} = 19.0 \text{ kN}$$

Note that in this instance of uniform width the result also can be obtained readily, of course, by computing the average pressure.

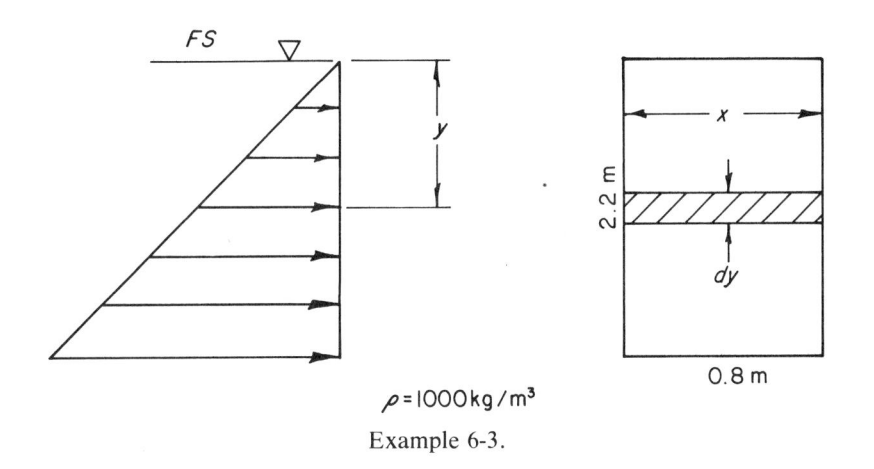

Example 6-3.

6.3 Buoyancy

The buoyancy force is the integrated vertical component of the forces exerted by the pressure of the liquid upon the contact surface of an immersed vessel. As usual the pressure is a function of $\rho g h$, i.e., it involves gravity. In accordance with the Archimedes Principle the buoyancy force is equal to the gravitational force on the displaced liquid and is equal and opposite to the gravitational force on the body.

Example 6-4

A vessel is partially immersed in a liquid of density 965 kg/m³ displacing a volume of 0.03 m³. What will be the buoyancy force on the vessel?

The resultant buoyancy force is:

(Eq. 6.3) $$F_b = V\rho g$$

$$F_b = 0.03(965)9.8 = 284 \text{ N}$$

$$m^3 \cdot \frac{kg}{m^3} \cdot \frac{m}{s^2} = \frac{kg}{s^2} \cdot m = N$$

Example 6-4.

6.4 Head

The pressure in a fluid is frequently expressed in terms of a head, h, which is, in effect, the depth in a fluid at which the corresponding hydrostatic pressure would be encountered. Since,

$$p = \rho g h \quad \text{in Pa;} \quad \text{then} \quad h = \frac{p}{\rho g} \quad \text{in meters}$$

Example 6-5

An inspection port in a pipeline carrying oil having a density of 850 kg/m³ is closed by a plate with an area of 0.16 m². What is the force on this plate when the head is 14 m?

$$p = 850(9.8)14 = 117 \text{ kPa}$$

$$\frac{\text{kg}}{\text{m}^3} \cdot \frac{\text{m}}{\text{s}^2} \cdot \text{m} = \frac{\text{kg} \cdot \text{m}}{\text{s}^2} \cdot \frac{1}{\text{m}^2} = \frac{\text{N}}{\text{m}^2} = \text{Pa}$$

$$F = pA = 117(0.16) = 18.7 \text{ kN}$$

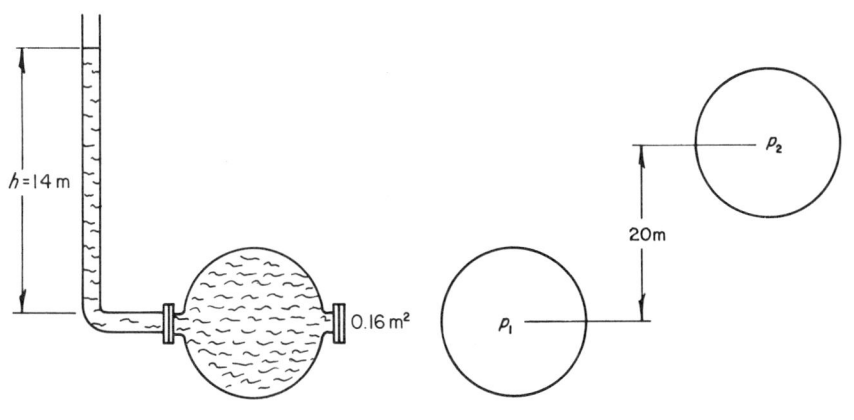

Example 6-5.   Example 6-6.

6.5 Center of Pressure

SI does not in any way change the customary calculations concerning the "center of pressure," the location of the resultant of pressure forces on a given surface, nor the calculations for the metacenter in buoyancy problems. Units would be in accordance with previous examples in Statics and in Hydrostatics.

6.6 Potential Energy

The potential energy of an element of fluid may consist of two components arising from: (1) pressure, due to confinement; (2) position, due to elevation above a reference datum. Like all other forms of energy, the potential energy of a fluid is expressed in joules, J.

For a given mass m of liquid, the potential energy associated with *pressure* is the product of pressure and volume, thus:

(Eq. 6.4a) $\quad E_p = pv \quad\quad \dfrac{N}{m^2} \cdot m^3 = N \cdot m = J$

(Eq. 6.4b) $\quad E_p = pm/\rho \quad\quad \dfrac{N}{m^2} \cdot kg \cdot \dfrac{m^3}{kg} = N \cdot m = J$

(Eq. 6.4c) $\quad E_p = mgh \quad\quad kg \cdot \dfrac{m}{s^2} \cdot m = N \cdot m = J$

In the special case of a unit volume, the potential energy is numerically equal to the pressure, but it should be kept in mind that the units of energy are always in joules, N·m. Hence, per unit volume:

(Eq. 6.5) $\quad e_p = \rho gh \quad\quad \dfrac{kg}{m^3} \cdot \dfrac{m}{s^2} \cdot m = N \cdot m/m^3 = J/m^3$

As previously established in the chapter on "Dynamics," the potential energy of a mass m due to position z above a given datum is:

(Eq. 6.6) $\quad E_z = mgz \quad\quad kg \cdot \dfrac{m}{s^2} \cdot m = N \cdot m = J$

Similarly, the potential energy, per unit of volume

(Eq. 6.7) $\quad e_z = \rho gz \quad\quad \dfrac{kg}{m^3} \cdot \dfrac{m}{s^2} \cdot m = N \cdot m/m^3 = J/m^3$

Thus a useful simplification for some computations of the combined potential energy per unit volume is:

(Eq. 6.8) $\quad e_p + e_z = \rho g(h + z) \quad\quad N \cdot m/m^3 = J/m^3$

in which h = pressure head and z = elevation.

Similarly, for a mass m the *total* potential energy is:

(Eq. 6.9) $\quad E_p + E_z = mg(h + z) \quad\quad N \cdot m = J$

Example 6-6

A 50 kg mass of liquid, having a density of 1030 kg/m³ in an initial position is under a pressure of 0.15 MPa. If this mass is moved to a position 20 m higher and the pressure is concurrently increased to 0.20 MPa, what is the resultant change in total potential energy?

Volume of fluid: $\quad 50/1030 = 0.0485 \text{ m}^3 \quad\quad kg \cdot \dfrac{m^3}{kg} = m^3$

Change in potential energy due to increase in pressure is pressure times volume:

$$p \cdot v = (0.20 - 0.15)0.0485 = 0.00243 \text{ MJ} = 2.43 \text{ kJ}$$

$$m^3 \cdot \frac{N}{m^2} = N \cdot m = J$$

and due to change in elevation is:

$$mgh = 50(9.8)20 = 9800 \text{ N} \cdot \text{m} = 9.8 \text{ kJ}$$

$$kg \cdot \frac{m}{s^2} \cdot m = N \cdot m = J$$

$$\therefore \text{ total gain in } E_p = 2.43 + 9.8 = 12.23 \text{ kJ}$$

If the approach were taken to express the energy in terms of "head," the numerical values would have been:

$$h_1 = \frac{p_1}{\rho g} = \frac{0.15 \times 10^6}{1.03 \times 10^3 (9.8)} = 14.86 \text{ m} \qquad \frac{kg \cdot m}{s^2} \cdot \frac{m^3}{kg} \cdot \frac{s^2}{m} = m$$

similarly,

$$h_2 = 19.81 \text{ m}$$

and

$$(h_2 - h_1) = 19.81 - 14.86 = 4.95 \text{ m}$$

Therefore, the combined change in potential energy is:

$$E_p = mg(h + z) = 50(9.8)(20 + 4.95) = 12.23 \text{ kJ}$$

6.7 Manometers

The "head" of liquid concept assumes a further significance when considering methods of measuring pressure. The simple U-tube manometer, and a variety of more complicated devices based upon it, operate by balancing an unknown pressure against a visible and measurable head. If the manometric liquid and the liquid whose pressure is being measured are one and the same, the correspondence between manometric head and actual head is obvious. If the liquids are different, the calculation of actual head from manometric head needs to account for the difference of the two densities. Clearly nothing new is involved here as regards units, but the computation will vary in accordance with the physical arrangement of the gages.

Example 6-7

Pressure readings on the suction and discharge lines of a pump are, respectively: -1.264 m of water; 1.697 m of mercury. The fluid being pumped is oil of specific gravity 0.85. The suction manometer tube connecting to the pump contains only air. On the discharge side, the reference point of the mercury gage is 0.630 m below the pressure tap. The vertical distance between the two pressure taps is 2.000 m. What is the total pressure head imparted to this fluid system by the pump? The suction and discharge lines are of the same diameter.

Pressure at A and at B, and thence the total head, may be determined by the application of Equation (6.1): $p = \rho g h$

$$p_A = -1000(9.807)1.264$$

$$= -12\,396 \text{ N/m}^2 \quad \frac{\text{kg}}{\text{m}^3} \cdot \frac{\text{m}}{\text{s}^2} \cdot \text{m} = \frac{\text{kg} \cdot \text{m}}{\text{s}^2} \cdot \frac{1}{\text{m}^2} = \frac{\text{N}}{\text{m}^2} = \text{Pa}$$

$$p_A = -12\,396 \text{ Pa}$$
$$p_A = -12.4 \text{ kPa}$$

Similarly:

$$p_C = 13\,570(9.807)1.697$$
$$= 225\,838 \text{ N/m}^2 = 225\,838 \text{ Pa} = 225.84 \text{ kPa}$$

$$(p_B - p_C) = 850(9.807)0.630$$
$$= 5252 \text{ N/m}^2 = 5252 \text{ Pa} = 5.25 \text{ kPa}$$

from which:

$$p_B = 225.84 - 5.25 = 220.59 \text{ kPa}$$

$$(p_B - p_A) = 850(9.807)2.000$$
$$= 16\,672 \text{ N/m}^2 = 16\,672 \text{ Pa} = 16.67 \text{ kPa}$$

Total Head $_{AB}$ to fluid $= 220.59 + 16.67 - (-12.40)$

$$= 249.66 \text{ kPa}$$

Alternately the pressures at A, B, and C and the respective differential pressures can be obtained (or checked) by the conversion factors cited in Section 6.2 as follows:

$p_A = -1.264$ m water $p_A = -1.264(9.807)$ $p_A = -12.40$ kPa

$p_C = 1.697$ m mercury $p_C = 1.697(133.08)$ $p_C = 225.84$ kPa

$p_B - p_C = 0.630$ m oil $p_B - p_C = 0.630(0.85)(9.807)$
$p_B - p_C = 5.25$ kPa $p_B\ \ \ \ \ \ \ = 225.84 - 5.25 = 220.59$ kPa

$z_B - z_A = 2.000$ m oil $2.000(0.85)(9.807) = 16.67$ kPa

Total Head $_{AB}$ to fluid $= 220.59 + 16.67 - (-12.40) = 249.66$ kPa

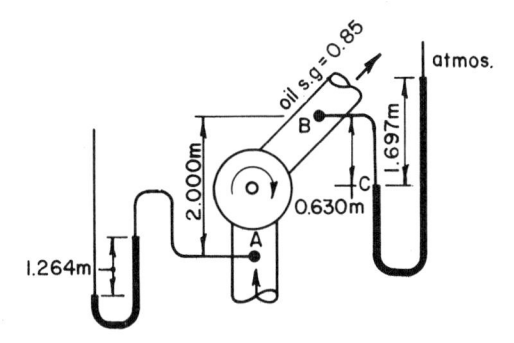

Example 6-7.

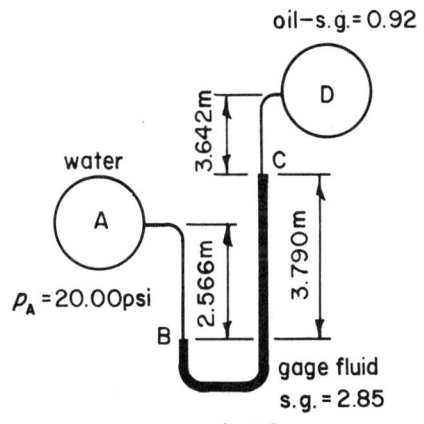

Example 6-8.

Example 6-8

A differential manometer using a gage fluid of specific gravity of 2.85, connects two pipelines A and D. Pipeline A contains water at a gage

156   FLUID MECHANICS   Ch. 6

pressure of 20.00 lbf/in.². What is the pressure in kPa in the oil at D when the position of the gage and the gage fluid is as indicated?

$$s_1 = 1.000 \text{ (water)} \quad s_2 = 0.920 \text{ (oil)} \quad s_g = 2.850 \text{ (gage)}$$
$$l_1 = 2.566 \text{ m}; \quad l_2 = 3.642 \text{ m}; \quad h = 3.790 \text{ m}$$

Noting that:
$$p_B = p_A + (\rho g s_1) l_2,$$
and that:
$$p_B = p_D + (\rho g s_2) l_2 + (\rho g s_g) h$$
whence:
$$p_D = p_A + (\rho g s_1) l_1 - (\rho g s_2) l_2 - (\rho g s_g) h$$

in which for water:

$$\rho g = 9807 \text{ Pa/m} \qquad \frac{\text{kg}}{\text{m}^3} \cdot \frac{\text{m}}{\text{s}^2} = \frac{\text{N}}{\text{m}^3} = \frac{\text{Pa}}{\text{m}}$$

also:
$$p_A = 20.00(6.8948) = 137.90 \text{ kPa (given)}$$

Solving for the pressure at D in kPa

$$p_D = 137.90 + 9.807(2.566) - 9.807(0.92)(3.642)$$
$$- 9.807(2.85)(3.790)$$
$$p_D = 24.27 \text{ kPa}$$

Alternately the solution may be viewed in these terms:

$$p_A = 20.00(6.8948) = 137.90 \text{ kPa}$$
$$p_B - p_A = 2.566(9.807) = 25.16 \text{ kPa}$$
$$p_B = 163.06 \text{ kPa}$$
$$p_C - p_B = 3.790(9.807)2.85 = -105.93 \text{ kPa}$$
$$p_C = 57.13 \text{ kPa}$$
$$p_D - p_C = 3.642(9.807)0.92 = -32.86 \text{ kPa}$$
$$p_D = 24.27 \text{ kPa}$$

6.8 Surface Tension

Manometers whose operation is based on the observation of liquid levels are subject to inaccuracy arising from meniscus effects, which are the consequence of surface tension. This phenomenon is defined as σ, the force per unit length and accordingly, the SI units, are N/m.

Example 6-9

A glass tube of 0.75 mm internal diameter is dipped vertically into water. How far will the water level rise in the tube?

The water rises because of "capillarity," which is a consequence of surface tension. It may be shown that the rise δ is a function of the diameter of the tube, the density, and the surface tension of the liquid in accordance with the following formula:

(Eq. 6.10) $\qquad \delta = \dfrac{4\sigma}{\rho g d} \qquad \dfrac{kg}{s^2} \cdot \dfrac{m^3}{kg} \cdot \dfrac{s^2}{m} \cdot \dfrac{1}{m} = m$

for water $\sigma = $ mN/m

$$\delta = \dfrac{4(73 \times 10^{-3})}{1000(9.8)(0.75 \times 10^{-3})}$$

$$\delta = 0.0397 \text{ m} = 39.7 \text{ mm}$$

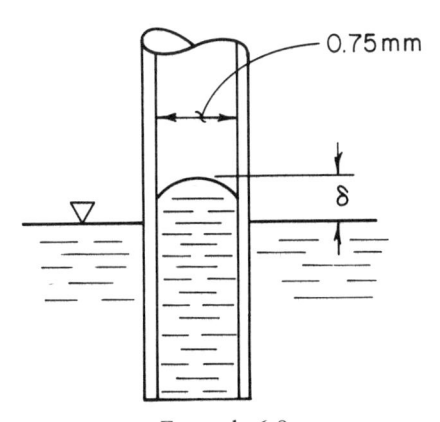

Example 6-9.

6.9 Incompressible Fluid Flow

The study of fluids in motion utilizes the concepts of velocity and acceleration which were introduced earlier in respect to solid bodies or particles under the general heading of Dynamics. Although in most fluid flows there is a variation in velocity across the stream, resulting in what is termed a "velocity profile," the general approach for most purposes is to utilize the "average velocity," usually denoted by $\bar{V}$. In terms of SI units $\bar{V}$ in m/s multiplied by the cross-section area to which it is applicable, A, usually given in m², results in a volumetric flow rate Q in m³/s. Thus:

(Eq. 6.11) $$Q = A\bar{V} \qquad m^2 \cdot \frac{m}{s} = m^3/s$$

The volumetric rate-of-flow multiplied by density of the fluid gives a mass rate-of-flow:

(Eq. 6.12) $$\dot{m} = Q\rho \qquad \frac{m^3}{s} \cdot \frac{kg}{m^3} = kg/s$$

Example 6-10

A liquid of density 950 kg/m³ flows at a rate of 15 kg/s through a pipe of 140 mm diameter. What is the volumetric rate-of-flow and the average velocity?

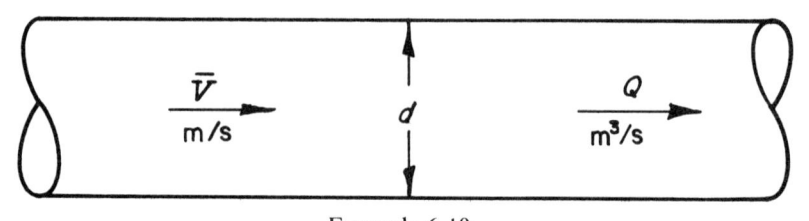

Example 6-10.

$$A = 0.7854(0.140)^2 = 0.0154 \text{ m}^2$$

$$Q = \frac{15}{950} = 0.0158 \text{ m}^3/s \qquad \frac{kg}{s} \cdot \frac{m^3}{kg} = m^3/s$$

$$\bar{V} = Q/A = \frac{0.0158}{0.0154} = 1.03 \text{ m/s} \qquad \frac{m^3}{s} \cdot \frac{1}{m^2} = m/s$$

Ch. 6   FLUID MECHANICS   159

6.10 Measuring the Rate-of-Flow

The most accurate way of measuring a flow rate is to make related measurements of quantity (i.e. volume or mass) and time; e.g., by collecting the fluid in a vessel, but this usually is not practicable. Indirect methods, such as the use of orifice plates or venturi meters, and other special devices such as rotameters and propeller meters, are commonly employed. These devices are usually calibrated against standards, or are exact copies of devices that have been calibrated. As already indicated, preferred units for rates of flow in SI are m³/s and kg/s, the use of the former being much more prevalent since mass flow meters are difficult to construct.

While the base units m³/s are used to describe most large flows the magnitudes of resultant quantities frequently are considered somewhat inconvenient for expression at low flows. For instance, with a velocity of 2 m/s, which is quite common for pipe flow, the delivery of 1.0 m³/s would involve a pipe approximately 800 mm in diameter.

There is, of course, nothing unusual in this situation and the unit prefixes are available to produce quantities of more convenient size. Therefore it is most common practice in respect to low flows to use the cubic decimeter which is 1/1000 of the cubic meter. The liter and the dm³ formerly were not identical, although the difference was extremely small. But since 1964 the liter has been redefined as exactly equal to the cubic decimeter and the two names now are synonymous. Thus, many small flows are expressed in liters per second (L/s). Examples of this will follow.

6.11 Energy Equation

Many problems in fluid mechanics are solved by the application of Bernoulli's Theorem. It will be useful to consider various forms expressed in SI units.

Essentially this is an "energy" equation which states that neglecting losses, for steady flow with fixed boundaries, the sum of the kinetic energy and the two potential energy components is constant. As would be expected each term of the equation is expressed in joules. For a mass m of fluid, at an elevation z above a given datum the equation is:

(Eq. 6.13) $\dfrac{m\bar{V}^2}{2} + mgz + \dfrac{mp}{\rho} = \text{constant}$ J

FLUID MECHANICS   Ch. 6

This expression in terms of SI units is:

$$(kg)\frac{m^2}{s^2} + (kg)\frac{m}{s^2} \cdot m + (kg)\frac{kg \cdot m}{s^2} \cdot \frac{1}{m^2} \cdot \frac{m^3}{kg} = N \cdot m = J$$

The equation frequently appears without the mass term m, implying energy *per unit mass*, the expression then being:

(Eq. 6.14) $\quad \dfrac{\bar{V}^2}{2} + gz + \dfrac{p}{\rho} = \text{constant/unit of mass} \qquad\qquad$ J/kg

In terms of SI this is:

$$\frac{m^2}{s^2} + \frac{m}{s^2} \cdot m + \frac{kg \cdot m}{s^2} \cdot \frac{1}{m^2} \cdot \frac{m^3}{kg} = \frac{m^2}{s^2}$$

but

$$\frac{m^2}{s^2} = \frac{kg \cdot m^2}{kg \cdot s^2} = \frac{N \cdot m}{kg} = \frac{J}{kg} = J/kg$$

If this equation is multiplied by the mass rate of flow in kg/s, or by the volumetric rate-of-flow and the density, (m³/s)(kg/m³), the various terms in the equation then represent rates of energy transfer, and are, accordingly, in joules per second, or watts.

The equation can also be expressed in terms of head of fluid; to achieve this the unit mass form is divided by g, whence the pressure term then becomes $p/\rho g$, which is denoted by h, and the equation takes what is probably the most frequently used form:

(Eq. 6.15) $\quad \dfrac{\bar{V}^2}{2g} + z + h = \text{constant} \qquad\qquad \dfrac{m^2}{s^2} \cdot \dfrac{s^2}{m} + m + m = m$

in which the unit of each term is the meter. To obtain the rate of energy transfer from this equation it is necessary to multiply by g and the mass flow rate, kg/s.

Example 6-11

A horizontal pipeline, carrying sea water, changes gradually in diameter from 1.5 m to 0.7 m. The pressure at the larger end is 30 kPa and the discharge flow rate is 1.2 m³/s. Neglecting losses, what will be the pressure at the smaller section? $\rho = 1030$ kg/m³.

Ch. 6   FLUID MECHANICS   161

Example 6-11.

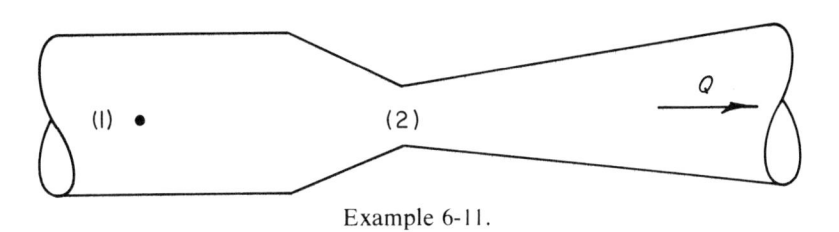

Applying the energy equation of Bernoulli in the form per unit of mass, namely, Equation (6.14):

$$\frac{\bar{V}^2}{2} + gz + \frac{p}{\rho} = \text{constant}$$

For horizontal pipe $z_1 = z_2$

$$A = \frac{\pi}{4}(1.5)^2 = 1.77 \text{ m}^2$$

$$\bar{V}_1 = \frac{Q}{A_1} = \frac{1.2}{1.77} = 0.678 \text{ m/s}$$

$$\bar{V}_2 = (0.678)\frac{(1.5)^2}{(0.7)^2} = 3.11 \text{ m/s}$$

$$\frac{(0.678)^2}{2} + gz_1 + \frac{30 \times 10^3}{1030} = \frac{(3.11)^2}{2} + gz_2 + \frac{p_2}{1030}$$

whence

$$p_2 = 30\,000 + 1030[(0.678)^2 - (3.11)^2]$$
$$= 30\,000 - 4744$$
$$p_2 = 25\,260 \text{ Pa} = 25.3 \text{ kPa}$$

Observe that, as always, SI units can be used directly in the formula without modification.

Example 6-12

Water flows through a pipe of gradually varying cross section. At a point where the diameter is 80 mm, the pressure head is 30 m; at another

162 FLUID MECHANICS Ch. 6

point, 5 m higher than the first, the diameter is 70 mm and the pressure head is 18 m. What is the volumetric rate-of-flow of water.

$$A_1 = \frac{\pi}{4}(0.08)^2 = 0.005026 \text{ m}^2$$

$$\bar{V}_1 = \frac{Q}{A_1} = \frac{Q}{0.005026} = 199Q$$

$$\bar{V}_2 = \bar{V}_1 \frac{(0.08)^2}{(0.07)^2} = 260Q$$

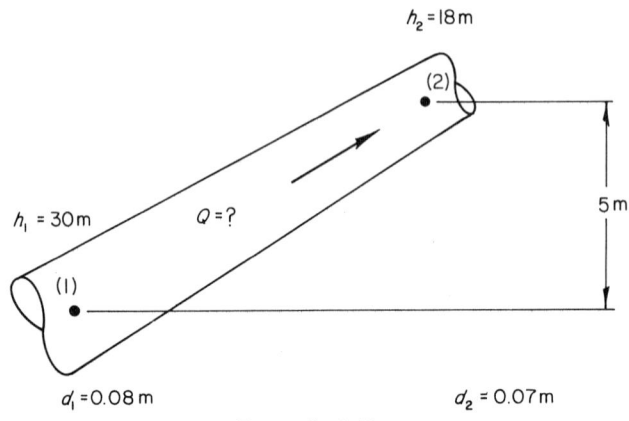

Example 6-12.

Using Bernoulli's energy equation expressed in terms of "head," in the form of Equation (6.15):

$$\frac{\bar{V}^2}{2g} + z + h = \text{constant}$$

$$\frac{(199Q)^2}{2(9.8)} + z_1 + 30 = \frac{260Q^2}{2(9.8)} + (z_1 + 5) + 18$$

whence

$$\frac{Q^2}{2(9.8)}\{(260)^2 - (199)^2\} = 7$$

$$Q = 0.07 \text{ m}^3/\text{s}$$

$$Q = 70 \text{ L/s}$$

Note that in this type of calculation, when using head consistently as a measure of pressure, the density does not appear.

6.12 Total Energy System

In most actual fluid flow systems there is a loss of energy, h_f, most frequently due to friction, or an input of energy, Δe_m, most often due to pumping, or some of each. The energy equations can be modified for use in such cases by algebraically adding, in the proper units, an energy gain and/or loss term; for example; Eq. 6.15 may be expanded as follows:

(Eq. 6.16) $$\frac{\bar{V}_1^2}{2g} + z_1 + h_1 - h_{f1-2} + \frac{\Delta e_m}{g} = \frac{\bar{V}_2^2}{2g} + z_2 + h_2$$

in which h_{f1-2} = head loss due to friction between two points (1) and (2); while $\Delta e_m/g$ is energy introduced (or withdrawn; use negative sign) between the same two points, expressed also in terms of "head" of fluid.

Example 6-13

Oil of density 930 kg/m³ is being pumped at a rate of 0.005 m³/s. The diameters of the suction and discharge pipes of the pump are respectively 50 mm and 75 mm; the suction and discharge gages read −40 kPa and 600 kPa, respectively, with the vertical distance between the pressure tappings being 8 m. Calculate the power required at the pump, assuming no pipe losses and a pump efficiency of 70 percent.

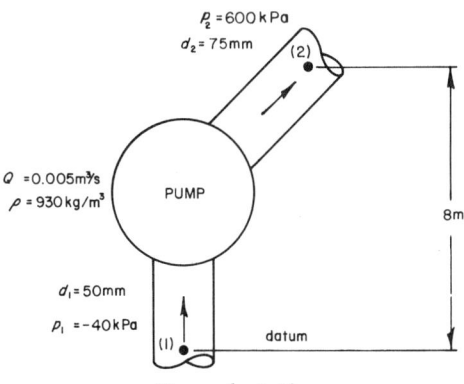

Example 6-13.

$$\bar{V}_1 = \frac{Q}{A_1} = \frac{0.005}{\frac{\pi}{4}(0.05)^2} = 2.55 \text{ m/s}$$

$$\bar{V}_2 = 2.55 \frac{(50)^2}{(75)^2} = 1.13 \text{ m/s}$$

Applying Equation (6.14), and adding Δe_m the energy per unit of mass supplied by the pump,

$$\frac{\bar{V}^2}{2} + gz + \frac{p}{\rho} = \text{constant} \quad \text{J/kg}$$

$$\frac{(2.55)^2}{2} + 0 - \frac{40 \times 10^3}{930} + \Delta e_m = \frac{(1.13)^2}{2} + 8(9.8) + \frac{600 \times 10^3}{930}$$

whence

$\Delta e_m = 764$ J/kg
$\dot{m}$ = the mass flow rate is: $0.005(930) = 4.65$ kg/s
P = power transmitted to the fluid is:

$$P = \Delta e_m \dot{m}$$

$$P = 764(4.65) = 3552 \text{ J/s} \qquad \frac{\text{J}}{\text{kg}} \cdot \frac{\text{kg}}{\text{s}} = \frac{\text{J}}{\text{s}} = \text{W}$$

The power consumed in pumping at 70 percent efficiency is:

$$3552/0.70 = 5074 \text{ J/s}, \quad \text{or} \quad 5.07 \text{ kW}$$

Note that in this example the energy supplied by the pump appears in Bernoulli's equation in the form of joules per kilogram, and has to be multiplied by a mass flow rate in kilograms per second to produce a power in watts. As already indicated, calculations may also be made in terms of head of fluid, expressed in meters, in which case the energy input term is also expressed in the form of a head. To obtain the power input in such instances the resultant energy value must be multiplied by the mass flow rate and by g, whence the units combine as follows:

(Eq. 6.17) $\qquad P = \dot{m}gh \qquad \frac{\text{kg}}{\text{s}} \cdot \frac{\text{m}}{\text{s}^2} \cdot \text{m} = \frac{\text{N} \cdot \text{m}}{\text{s}} = \frac{\text{J}}{\text{s}} = \text{W}$

6.13 Friction Head

In a similar manner, terms representing losses due to fluid friction may also be represented as energy, energy per unit mass, or head—depending on the form of the equation found most convenient.

Example 6-14

Water flows at 5.0 m³/s through a straight pipe of uniform diameter at a rate which produces a loss of head through friction of 5 mm per meter

Ch. 6   FLUID MECHANICS   165

length. Find the difference in pressure between two points at 100 m distance apart, if the downstream point is 8 m higher than the upstream point.

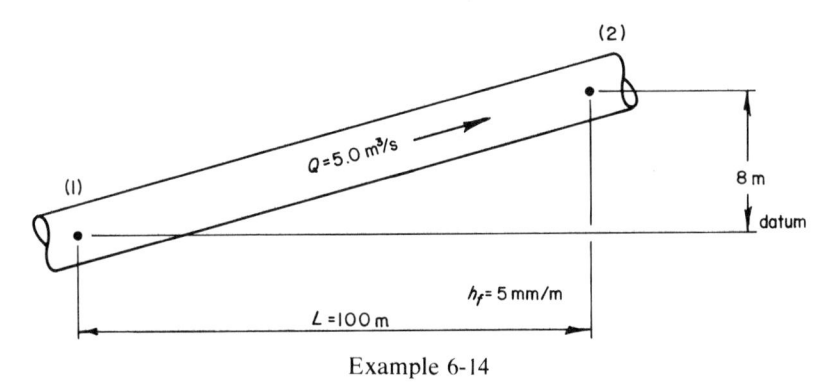

Example 6-14

Total loss of "head" due to friction is: $100(0.005) = 0.5$ m. Since the pipe is of constant diameter Bernoulli's equation in "head" form, Equation (6.15), reduces to:

$$z_1 + h_1 - h_{fl-2} = z_2 + h_2$$
$$0 + h_1 - 0.5 = 8 + h_2$$

Hence, the pressure difference expressed in terms of head is:

$$h_1 - h_2 = 8.5 \text{ m}$$

Expressed in unit pressure terms this becomes:

$p = \rho g h$

$p = 1000(9.8)8.5 = 83\,300$ Pa, or, 83.3 kPa

The corresponding power consumption, P, is obtained either: (1) by multiplying the head loss in meters by g and by the mass flow rate in kg/s, or (2) by multiplying the pressure difference in pascals by the volumetric flow rate in m³/s.

First Method: (from Equation [6.4c])

(Eq. 6.18a) $P = \dfrac{E}{t} = \dfrac{m}{t}(gh) = mgh/t$

$P = 5.0(1000)9.8(8.5) = 416\,500$ W $= 416$ kW

$$\dfrac{m^3}{s} \cdot \dfrac{kg}{m^3} \cdot \dfrac{m}{s^2} \cdot m = \dfrac{kg \cdot m}{s^2} \cdot \dfrac{m}{s} = \dfrac{N \cdot m}{s} = \dfrac{J}{s} = W$$

Second Method: (from Equation [6.4b])

(Eq. 6.18b) $$P = \frac{E}{t} = (p)\frac{m}{\rho t} = (p)\dot{m}/\rho$$

$$kg \cdot \frac{m}{s^2} \cdot \frac{1}{m^2} \cdot \frac{kg}{s} \cdot \frac{m^3}{kg} = \frac{N \cdot m}{s} = \frac{J}{s} = W$$

$$P = 83\,300(5.0) = 416\,500 \text{ W} = 416 \text{ kW}$$

This type of calculation has been considered at some length because it provides another instance of the changed role of g when using SI units. The use of head as an indicator of pressure is based on the variation of pressure with height which results from the effect of gravity. To convert a head into a true pressure it is necessary to introduce g as a factor.

On the other hand, in the basic Bernoulli equation the kinetic energy term, $m\tilde{V}^2/2$, has nothing to do with gravity and accordingly, g does *not* appear. Conversely, when working in terms of head, the kinetic energy term has to be divided by g, whereas the head term is left as it is.

Note once again that the derivation of power from energy involves no numerical factors; the joule per second is the same as the watt.

6.14 Flow Meter

The venturi meter uses the change in pressure head accompanying the acceleration of a fluid as a means of observing flow rate.

Example 6-15

A venturi meter inserted into a 120-mm-diameter pipe conveying water has a throat diameter of 80 mm. The pressure difference between full pipe and throat is determined to be 550 mm of mercury.

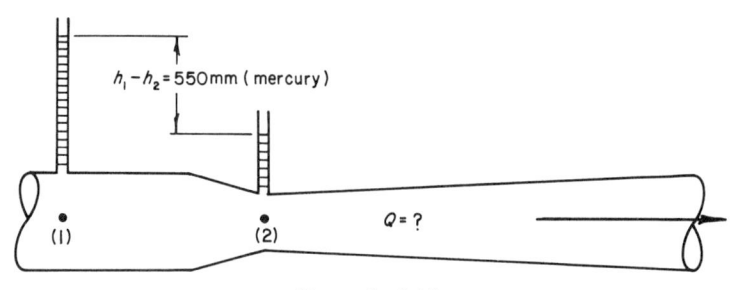

Example 6-15.

Assuming no losses, what is the indicated flow rate? Head across the venturi is:

$$0.55(13.57) = 7.48 \text{ m (water)} = h_1 - h_2$$

Applying the energy Equation (6.15):

$$\frac{\bar{V}_1^2}{2g} + z_1 + h_1 = \frac{\bar{V}_2^2}{2g} + z_2 + h_2$$

noting that

$$z_1 = z_2$$

and

$$h_2 = h_1 - 7.48$$

$$\frac{Q^2}{[(0.12)^2 \pi/4]^2 \, 2g} + z_1 + h_1 = \frac{Q^2}{[(0.08)^2 \pi/4]^2 \, 2g} + z_2 + h_1 - 7.48;$$

whence

$$Q = 0.0674 \text{ m}^3/\text{s}$$

In practice such a calculation probably would be based on a formula which by-passes the reasoning involved, such as:

(Eq. 6.19) $\qquad Q = \dfrac{c_d A_2 \sqrt{2g \Delta h}}{\sqrt{1 - (A_2/A_1)^2}} \quad$ or, $\quad Q = K\sqrt{\Delta h}$

where A_1 and A_2 are the areas of pipe and throat, respectively, Δh is the differential head, C_d is an empirically determined coefficient of discharge which allows for losses. The simplified calculation above, assumed that $C_d = 1.0$. This procedure, of course, gives the same result in units:

$$\text{m}^2 \left\{ \frac{\text{m}}{\text{s}^2} \cdot \text{m} \right\}^{1/2} = \frac{\text{m}^3}{\text{s}}$$

Note that if the $\sqrt{2g}$ is extracted from an equation such as this, the value in SI units is $\sqrt{2(9.81)} = 4.43$, which, of course, is quite different from the comparable value in customary units.

6.15 Force of a Jet

The discussion of Bernoulli's Theorem and related matters introduced the concept of the kinetic energy of a moving fluid. The concept of

168   FLUID MECHANICS   Ch. 6

momentum is equally straightforward, as the following example will probably be sufficient to demonstrate.

Newton's Law may be expressed as: "Force is proportional to rate-of-change of momentum." With SI units the coefficient of proportionality is unity.

(Eq. 6.20) $F = \dot{m}\Delta V$

$$F = Q\rho\Delta V \qquad \frac{m^3}{s} \cdot \frac{kg}{m^3} \cdot \frac{m}{s} = \frac{kg \cdot m}{s^2} = N$$

Note that g does not appear in this equation, the process being independent of gravity, and that the unit for rate-of-change of momentum is the newton.

Example 6-16

A jet of water, 30 mm in diameter, issues from a nozzle at an average speed of 20 m/s, and impinges normal to a surface which reduces it to rest. What force is exerted on the surface?

In one second the volume of water reduced to rest is:

$$(\pi/4)(0.03)^2 \quad 20 = 0.01414 \text{ m}^3$$

The corresponding mass is 14.14 kg
The momentum of this mass is $(14.14)(20) = 282.8$ kg·m/s
Since this amount of momentum is dissipated in one second the rate-of-change of momentum is 282.8 $\dfrac{kg \cdot m}{s^2}$

Accordingly, the force is 282.8 N

6.16 Pitot Tube

A pitot tube is a device for measuring velocity in a fluid by bringing an element of flow to rest, thus converting velocity head to additional static head.

A pitot-static device measures velocity head in respect to ambient pressure.

Example 6-17

The velocity of air in a duct at atmospheric pressure is measured by a pitot-static tube.

The observed differential pressure is 80 mm of water. What is the measured velocity? Since, by Equation (6.1):

$$p = \rho g h \qquad p = 1000(9.8)0.08 = 784 \text{ Pa}$$

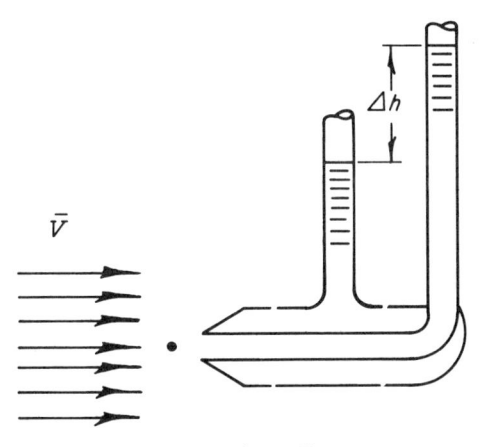

Example 6-17.

from Bernoulli's Equation (6.14) in terms of J/kg:

$$\frac{\bar{V}_1^2}{2} + gz_1 + \frac{p_1}{\rho} = 0 + gz_2 + \frac{p_1 + 784}{\rho}$$

taking $\rho = 1.23$ kg/m³ for air at atmospheric pressure

$$\frac{\bar{V}_1^2}{2} = \frac{784}{1.23} \qquad \bar{V}_1 = 35.7 \text{ m/s}$$

As in the case of the venturi meter, a simplified formula including a calibration or correction factor would normally be used for this calculation.

6.17 Viscosity

Calculations concerning the loss of head arising from fluid friction introduce the concept of viscosity, a physical property representing the inter-molecular forces which accompany a variation in velocity between adjoining layers of a moving fluid, as shown in the accompanying sketch. A shearing stress, τ, in a fluid establishes a rate of shearing strain usually considered to vary uniformly over small distances in the velocity profile (see Fig. 6-3). The factor of proportionality is known as

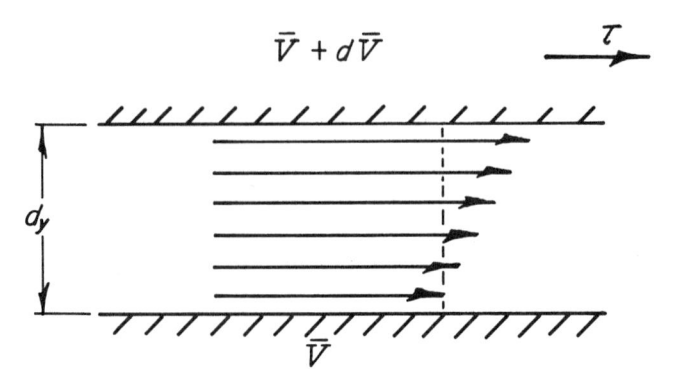

Fig. 6-3. Velocity profile.

the viscosity of the fluid, μ. Checking units:

$$\mu = \frac{\text{shear stress}}{\text{rate of shear strain}} = \frac{\tau}{\frac{dV}{dy}} \qquad \frac{N/m^2}{\left(\frac{m}{s}\right) \cdot \frac{1}{m}}$$

$$\mu = \text{Pa} \cdot \text{s}$$

This is the dynamic viscosity and the SI unit is known as the "pascal-second" (Pa·s). Typical values are given in Table 6-3.

In some instances it is desirable to express μ in terms of SI base units, thus:

$$\mu = \text{Pa} \cdot \text{s} = \frac{N}{m^2} \cdot \text{s} = \text{kg} \cdot \frac{m}{s^2} \cdot \frac{1}{m^2} \cdot \text{s}$$

Hence, alternately:

$$\mu = \frac{\text{kg}}{m \cdot s}$$

Many calculations involve both the viscosity and the density of the fluid. As a matter of convenience these are combined in a term known as the kinematic viscosity, ν. The applicable units are:

$$\nu = \frac{\mu}{\rho} = \frac{\text{Pa} \cdot \text{s}}{\text{kg}/m^3} = \frac{\text{kg} \cdot m}{s^2} \cdot \frac{1}{m^2} \cdot \text{s} \cdot \frac{m^3}{\text{kg}} = \frac{m^2}{s}$$

which is called the "meter squared per second."

Ch. 6 FLUID MECHANICS 171

Since earlier metric units for viscosity are well established in industry and science, it may be well at this point to recall the units used and indicate the relationships to SI. The metric unit for dynamic viscosity (μ) has been the "poise" (P) which is defined as:

$$P = \frac{\text{gram}}{\text{cm-sec}}$$

$$P = \frac{0.001 \text{ kg}}{0.01 \text{ m·s}} = 0.1 \text{ kg} \cdot \frac{\text{m}}{\text{s}^2} \cdot \frac{1}{\text{m}^2} \cdot \text{s} = 0.1 \text{ Pa·s}$$

Obviously, the centipoise (cP, or, Z) is equal to 0.001 Pa·s = 1.0 mPa·s. For kinematic viscosity (ν) the metric unit has been the "stoke" which is defined as:

$$\text{St} = \frac{\text{cm}^2}{\text{sec}} \qquad \text{St} = \frac{(0.01 \text{ m})^2}{\text{s}} = 10^{-4} \frac{\text{m}^2}{\text{s}}$$

Obviously, the centistoke (cSt) is equal to 10^{-6} m²/s, or, mm²/s. Another essential relationship in the expression of viscosity is to that given in customary units, which is found as follows:

$$1.0 \text{ P} = \frac{\text{gram(mass)}}{\text{cm-sec}} \left(\frac{1}{453.8} \frac{\text{lb}}{\text{g}}\right) 2.54 \frac{\text{cm}}{\text{in.}} \left(\frac{1}{32.17(12)} \frac{\text{sec}^2}{\text{in.}}\right)$$

$$= 1.4499 \times 10^{-5} \frac{\text{lbf·sec}}{\text{in.}^2}$$

$$1.0 \text{ P} = 1.45 \times 10^{-5} \frac{\text{lbf·sec}}{\text{in.}^2}$$

$$69\,000 \text{ P} = 1.0 \frac{\text{lbf·sec}}{\text{in.}^2} = 1.0 \text{ *Reyn} = \mu$$

6.18 Pipe Flow

A common engineering problem is that of loss of head due to friction in straight pipes of circular cross section. The basis for a brief review of this phenomena, involving the factors of pressures and of viscous flow is

---

* This unit may also be found expressed in terms of lbf·sec/ft², in which further unit adjustment will need to be made. "Reyn" is a title sometimes used in recognition of Osborne Reynolds, a well-known scientist. For further discussion of these units see "Conversion to Preferred SI Usage," in Chapter 9.

172 FLUID MECHANICS Ch. 6

given in Fig. 6-4. Such analysis also provides an excellent opportunity to illustrate key SI unit relationships.

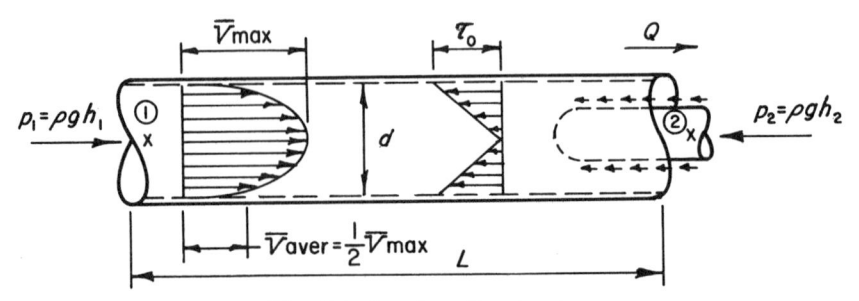

Fig. 6-4. Laminar flow in pipe.

Assuming that flow is laminar, consisting of a series of concentric fluid cylinders sliding one within the other, the shear stress τ will vary as indicated and the variation in velocity will be in paraboloid form as shown. From this analysis the following theoretical relationship is derived.

(Eq. 6.21) $$\frac{p_1 - p_2}{\rho g} = h_{f1-2} = \frac{32 \mu \bar{V} L}{\rho g \, d^2} = \frac{32 \nu \bar{V} L}{g \, d^2}$$

This expression is known as Poiseuille's equation and applies in this form to laminar flow only. See discussion of Reynolds Number which follows. Also note that $\bar{V}$ is the average velocity obtained from $\bar{V} = Q/A$.

6.19 Reynolds Number

A generally useful indicator of flow characteristic is found from the ratio of inertial forces to viscous forces in the form known as Reynolds Number (R), namely:

(Eq. 6.22) $$R = \frac{\bar{V} d \rho}{\mu}, \quad \text{or,} \quad R = \frac{\bar{V} d}{\nu}$$

for which the unit check-out of ($\bar{V} d\rho/\mu$) gives the following:

$$\frac{(m/s)(m)(kg/m^3)}{Pa \cdot s} = \frac{m}{s} \cdot m \cdot \frac{kg}{m^3} \cdot \frac{m \cdot s^2}{kg \cdot s} = 1 = \text{neutral dimensions}$$

A ratio such as R is frequently referred to as a dimensionless number but it seems more factual to refer to a value such as R as being of "neutral dimensions." The expression ($\bar{V} d/\nu$) would check out similarly.

Laboratory studies show that for values of $R < 2000$, flow is laminar and Equation (6.21) applies. For higher values of R, flow is in transition, or is turbulent, and friction factors are obtained from compilations of experimental data.

For the conditions of laminar flow, a key relationship—confirmed by experimental data—may be obtained from Equation (6.21) as follows:

$$h_{f1-2} = \frac{32 \nu \bar{V} L}{g d^2} = 64 \cdot \frac{\nu}{\bar{V} d} \cdot \frac{L}{d} \cdot \frac{\bar{V}^2}{2g} = f \frac{L}{d} \frac{\bar{V}^2}{2g}$$

from which

(Eq. 6.23) $$f = \frac{64}{R}$$

6.20 Friction Factor

Another general approach may be obtained by developing a relationship between overall viscous forces and inertial forces in the pipe.

The total friction force to be overcome is: $\tau \pi \, dL$

The total inertial forces lost in friction are a function of: $\frac{1}{2}\rho \bar{V}^2$. The factor of proportionality is known as the friction factor, f', which combines the effect of several variables, thus:

$$(p_1 - p_2)\frac{\pi d^2}{4} = \tau \pi \, dL \quad \text{but since} \quad \tau = f'\rho \frac{\bar{V}^2}{2},$$

it may be written:

(Eq. 6.24a) $$(p_1 - p_2) = 4f'\rho \frac{\bar{V}^2}{2} \frac{L}{d} = 2f'\rho \bar{V}^2 \frac{L}{d}$$

This expression is sometimes re-written as:

(Eq. 6.24b) $$p_1 - p_2 = \tfrac{1}{2} f \rho \bar{V}^2 L / d$$

in which $f = 4 f'$ (see discussion below).

Further modification of Equation (6.24a) yields:

$$\frac{p_1 - p_2}{\rho g} = 4 f' \cdot \frac{L}{d} \cdot \frac{\bar{V}^2}{2g}$$

which in turn is usually written:

(Eq. 6.25a) $$h_{f1-2} = f \frac{L}{d} \frac{\bar{V}^2}{2g}$$

giving the expression generally known as the Darcy-Weisbach equation. The Darcy f is common in hydraulics. The factor f' (sometimes called the Fanning factor) is more frequently used in heat transfer and aerodynamics, and as will be noted, is one fourth as large. Values of f and f' are available from various sources of applicable experimental data.

It is a valuable feature of the International System that data in SI units may be inserted directly into appropriate formulae without the need for introducing conversion factors of any kind. This follows directly from the principle of coherence on which the system is based. An illustration of this characteristic appears again in the basic expression for f and f' where insertion of the appropriate SI units gives the following results in terms of units:

$$f = 4f' = \frac{\tau}{\tfrac{1}{2}\rho \bar{V}^2} \qquad \frac{N/m^2}{(kg/m^3)(m/s)^2} = \frac{kg}{m \cdot s^2} \cdot \frac{m^3}{kg} \cdot \frac{s^2}{m^2} = 1$$

This verifies the neutral dimension nature or "ratio" characteristic of both of these factors, i.e., such a factor is dimensionally independent, it is a universal factor not dependent in any way on the unit system being used as long as a single unit system is applied consistently throughout any particular use.

6.21 Pipe Flow Calculations

Many calculations concerning the flow of fluids are based on formulae, charts, and diagrams which present experimental data in a condensed form by utilizing the results of dimensional analysis to determine values of f. An example is the general expression for the frictional resistance due to turbulent flow in a straight pipe of circular cross section:

$$f \sim \frac{\tau}{\tfrac{1}{2}\rho \bar{V}^2} \sim \left(\frac{\bar{V} d \rho}{\mu}\right)^n$$

here τ is the frictional force per unit area of pipe surface, ρ the density of the fluid, $\bar{V}$ the average velocity, d the diameter of the pipe, and μ the dynamic viscosity. The result of applying dimensional analysis is to present data covering five distinct parameters τ, ρ, $\bar{V}$, d, and μ, in terms of two groups, is to give a direct relationship between the friction factor f and the Reynolds number R. A typical resultant formula is one generally known as the Blasius expression for the friction in smooth pipes:

(Eq. 6.25b) $\qquad f = 0.316/R^{0.25}$

This expression has been shown to be valid up to values of $R = 100\,000$. Beyond that value other formulae or charts should be consulted.

Example 6-18

A liquid of specific gravity 0.9 and kinematic viscosity of 20 cSt is pumped at 0.3×10^{-3} m³/s through a tube with a diameter of 12 mm. Find the pressure drop over a length of 5 m.

This example introduces terms and units which are obsolescent. Specific gravity means, in this case, the density of the fluid divided by the density of water; a more correct term would be "relative density." Here it is implied that the density is $0.9(1000) = 900$ kg/m³. The centistoke is: 10^{-6} m²/s, so that the kinematic viscosity is 20×10^{-6} m²/s. The dynamic viscosity is:

$$\mu = \nu\rho = 20 \times 10^{-6}(900) = 0.018 \text{ Pa·s}$$

$$\frac{m^2}{s} \cdot \frac{kg}{m^3} = \frac{kg \cdot m}{s^2} \cdot \frac{1}{m^2} \cdot s = \frac{N}{m^2} \cdot s = Pa \cdot s$$

The mean velocity $= \dfrac{0.3 \times 10^{-3}}{(0.012)^2 \pi/4} = 2.65$ m/s

and the Reynolds number $R = \bar{V} d/\nu = 2.65 \times 0.012/20 \times 10^{-6} = 1591$. Since $R = 1591 < 2000$, laminar flow is indicated; thus $f = 64/R$ whence $f = 0.0402$. From Equation (6.24b)

$$(p_1 - p_2) = \tfrac{1}{2} f \rho \bar{V}^2 L/d$$

$$= \tfrac{1}{2}(0.0402)900(2.65)^2(5)/0.012$$

$$= 52\,930 \text{ Pa}$$

$$= 52.9 \text{ kPa}$$

A great variety of formulae and charts for estimating pressure losses in pipes involve units in the same way. Examples of these follow. Note that the factors given must be applied with a full appreciation of the nature and scope of the original data and the specific units involved.

Example 6-19

A liquid of mass density $\rho = 910$ kg/m^3 and dynamic viscosity 1.2 mPa·s is pumped through a pipe 600 m in length and 80 mm in diameter at a rate of 4 kg/s. Find the pressure drop and the power consumed.

$$\text{Vol. flow rate} = 4/910 = 0.004396 \text{ m}^3/\text{s}$$

$$\text{Mean velocity} = \frac{0.004396}{(0.08)^2 \pi/4} = 0.8745 \text{ m/s}$$

From Equation (6.22), Reynolds number is:

$$R = \frac{\bar{V} d \rho}{\mu}$$

$$R = \frac{(0.8745)(0.08)(910)}{1.2 \times 10^{-3}}$$

$$R = 53\,050 > 4000 \quad \therefore \text{ Turbulent flow will occur}$$

Since $R < 100\,000$, Equation (6.25b) may be used:

$$f = 0.316/(53\,050)^{0.25}$$

$$f = 0.0208$$

from which

$$f' = f/4 = 0.0052$$

Applying Equation (6.24a)

$$p_1 - p_2 = f' \rho \bar{V}^2 \frac{L}{d} = 2(0.0052)(910)(0.8745)^2(600)/0.08 = 54\,280 \text{ Pa}$$

The power consumed is the product of the pressure drop and the volumetric flow rate:

$$0.004396(54\,280) = 240 \text{ W}$$

Admittedly several short cuts might have been taken in the example such as using ν to obtain R and applying f directly in an alternate expression, but the above route was chosen to show more fully the tie-in with basic expressions as found in the literature.

Example 6-20

What is the lost head to be expected in a cast-iron pipe 200 mm in diameter and 610 m long which is discharging 0.085 m³/s of water at 20°C for which $\nu = 1.69 \times 10^{-6}$ m²/s.

A = cross-sectional area of pipe = $(0.2)^2 \pi/4 = 0.0314$ m²

$\bar{V}$ = velocity in pipe = $0.085/0.0314 = 2.71$ m/s

$R = \bar{V} d/\nu = 2.71(0.2)/1.69 \times 10^{-6}$ $R = 320\,000$

as shown in Fig. 6.5,[1] $f = 0.0225$. Using Equation (6.25a):

$h_f = 0.0225(610/0.2)(2.71)^2/2(9.8) = 25.71$ m of water

$$\frac{m}{m} \cdot \frac{m^2}{s^2} \cdot \frac{s^2}{m} = m$$

6.22 Other Pipeline Losses

The loss of head caused by pipeline fittings, valves, bends, etc., is usually calculated by multiplying the velocity head by an experimentally

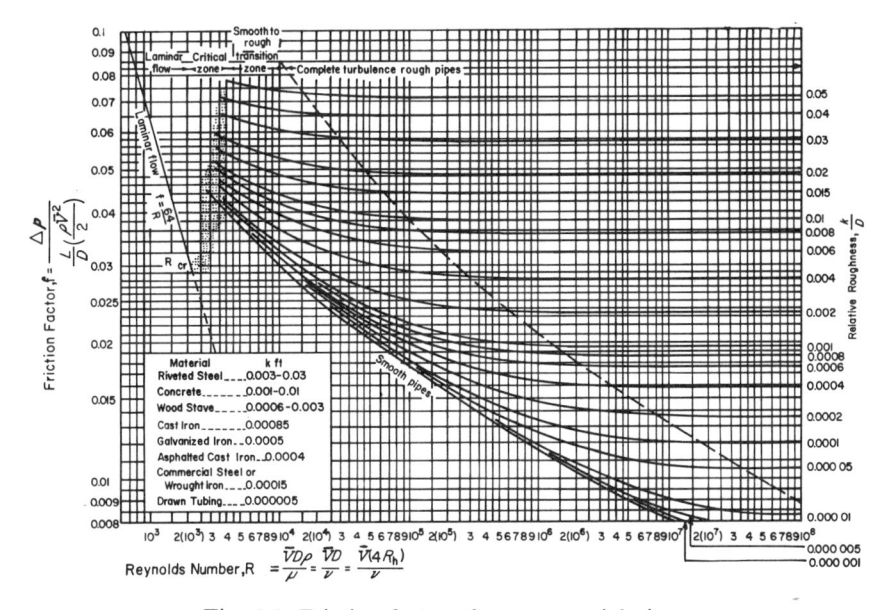

Fig. 6-5. Friction factors for commercial pipe.

[1] L. F. Moody, trans. "Friction Factors for Pipe Flow," *ASME*, Vol. 66, 1944.

determined coefficient, K_1. Thus:

$$h_L = K_1 \frac{\bar{V}^2}{2g} \qquad \frac{m^2}{s^2} \cdot \frac{s^2}{m} = m$$

The equivalent pressure drop for the same fluid is obtained by multiplying by ρg.

Example 6-21

If the coefficient for loss of head at a valve in the pipeline of Example 6-20 is estimated to be given by $K_1 = 0.25$, in terms of velocity head, what is the pressure drop across this valve?

$$\bar{V} = 2.71 \text{ m/s} \qquad \bar{V}^2/2g = (2.71)^2/2(9.8) = 0.375 \text{ m}$$

thus, the loss of head is: $0.375(0.25) = 0.0938$ m of water and the pressure drop is given by Equation (6.1):

$$p = \rho g h$$
$$1000(9.8)0.0938 = 0.919 \text{ kPa}$$

$$\frac{kg}{m^3} \cdot \frac{m}{s^2} \cdot m = \frac{kg \cdot m}{s^2} \cdot \frac{1}{m^2} = \frac{N}{m^2} = Pa$$

6.23 Lubrication of Bearings

The design of bearings is concerned with the maintenance of a film of lubricant between the bearing surfaces so that it can withstand the applied load without exerting too much drag. It is, accordingly, much concerned with force, pressure, and viscosity—quantities which have already been encountered.

Example 6-22

A simple thrust bearing consists of a flat rectangular pad, 200 mm long and 50 mm wide, moving in a direction parallel to its longer edges at a speed of 4 m/s on a large, flat surface. It is lubricated with a fluid having a dynamic viscosity of 45 mPa·s. The clearance of the trailing edge is 0.07 mm. What is the indicated load which the bearing is supporting?

One of the typical metric practice expressions for the average pressure in the lubricant film is:

$$p = \frac{6\mu u L}{h^2} \cdot K$$

μ = dynamic viscosity   Pa·s
p = lubricant pressure   Pa
h = film thickness at trailing edge   m
L = length of pad   m
u = relative speed   m/s

K is a coefficient depending upon the angle between the two surfaces being lubricated. A probable value for the conditions of this problem is $K = 0.0265$.

Then:

$$p = \left[\frac{6(45 \times 10^{-3})4(0.05)}{(0.07 \times 10^{-3})^2}\right] 0.0265 \qquad p = 292 \times 10^3 \text{ Pa}$$

Total force on bearing is: $292 \times 10^3(0.05)0.200 = 2920$ N $= 2.92$ kN. This would correspond to the weight of a mass of $2920/9.8 = 298$ kg.

Example 6-23

A simple journal bearing consists of an axle 50 mm in diameter in a bearing 75 mm in length, with a clearance modulus of 0.002 mm/mm. The viscosity of the lubricant is 34 centistokes. Under no-load conditions, what torque will be required to rotate the axle at 500 r/min? What power consumption does this represent? Assume an appropriate equation from metric practice is:

$$T = \mu r A \bar{V}/h$$

T = applied torque   N·m/rad
μ = dynamic viscosity   Pa·s
r = radius of bearing   m
A = bearing area   m²
$\bar{V}$ = relative linear speed   m/s
h = radial clearance   m

Since the kinematic viscosity has been given this must be converted to dynamic viscosity. Assume the density of the lubricant to be 880 kg/m³.
Note that

34 cSt = 34×10^{-6} m²/s

then

$$\mu = 34 \times 10^{-6}(880) = 0.030 \text{ Pa·s} \qquad \frac{m^2}{s} \cdot \frac{kg}{m^3} = \frac{kg}{m \cdot s} = \text{Pa·s}$$

$$A = \pi(0.05)0.075 = 0.01178 \text{ m}^2$$

$$\tilde{V} = \pi(0.05)500/60 = 1.31 \text{ m/s}$$

$$h = (0.002)25 = 0.050 \text{ mm} = 0.000\,05 \text{ m}$$

from whence

$$T = 0.030(0.025)0.01178(1.31)/0.000\,05$$

$$T = 0.23 \text{ N·m/rad}$$

$$\frac{kg}{m \cdot s} \cdot \frac{m}{rad} \cdot m^2 \cdot \frac{m}{s} \cdot \frac{1}{m}$$

$$= \frac{kg \cdot m}{s^2} \cdot m = \frac{N \cdot m}{rad} = \text{J/rad}$$

The power consumption is found, as usual, by multiplying by the angular velocity in radians/second:

i.e., $(500)2\pi/60 = 52.4$ rad/s from which $P = 0.23(52.4) = 12$ W

6.24 Comparison with Customary Practice

In order to provide an approximate comparison with customary practice, an analysis may be made of a bearing similar to that in Example 6-23, using Machinery's Handbook (20th ed.) as a reference.[2]

It will be noted that in this reference, as in many others, direct use is made of the centistoke in various formulas, although customary units have been used for all other dimensions. As a result there is a "mix" of units in the formulas given which necessitates the introduction of conversion factors such as the 1.45×10^{-5} and 69 000 discussed earlier in this chapter under the title of "Viscosity."

In proceeding with such a comparison it will be found expeditious to individually provide temporary auxiliary scales on open margins of present reference charts having customary units only.

For instance, Fig. 6-6, originally printed only in the customary units, now has millimeter scales added on the top and right-hand margins.

---

[2] E. Oberg and F. D. Jones. *Machinery's Handbook, 20th ed.*, New York: Industrial Press Inc., 1975.

Ch. 6 FLUID MECHANICS 181

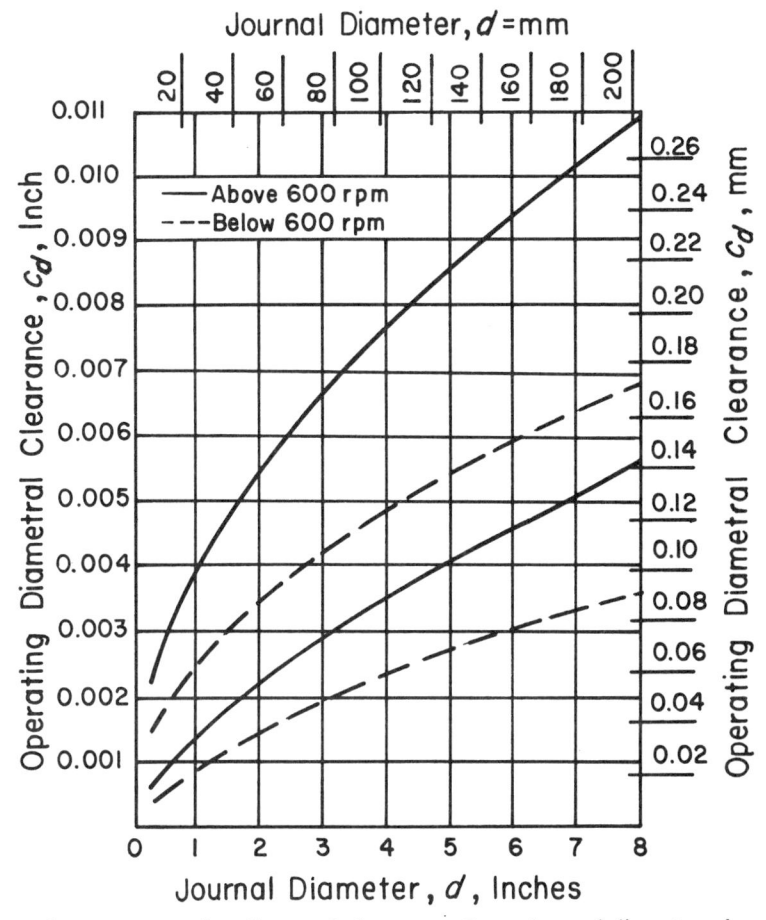

Fig. 6-6. Operating diametral clearance, C_d vs. journal diameter, d.

Until printed metric charts are generally available this temporary mode of reference will be found quite serviceable. Conversion scales on such charts can be readily developed by identifying two common reference points and then sub-dividing by a method of proportioning.

A demonstration of this form of approach to SI design follows.

Example 6-24

A simple journal bearing consists of a shaft of 50 mm diameter in a bearing of 75 mm length. The viscosity of the lubricant is 34 centistokes.

Table 6-4. Allowable Sleeve Bearing Pressures for Various Classes of Bearings*

Type of Bearing or Kind of Service	Pressure kPa = kN/m²
Electric Motor and Generator Bearings (general)	700–1400
Turbine and Reduction Gears	700–1750
Heavy Line Shafting	700–1050
Light Line Shafting Automotive	100–250
Main Bearings	3500–4500
Rod Bearings	10 000–17 500
Centrifugal Pumps	550–700

* These pressures in kilopascals (or kilonewtons per square meter) are based on area equal to length times diameter and are intended as a general guide only. The allowable unit pressure depends on operating conditions, especially in regard to lubrication, design of bearings, workmanship, velocity, and nature of load.

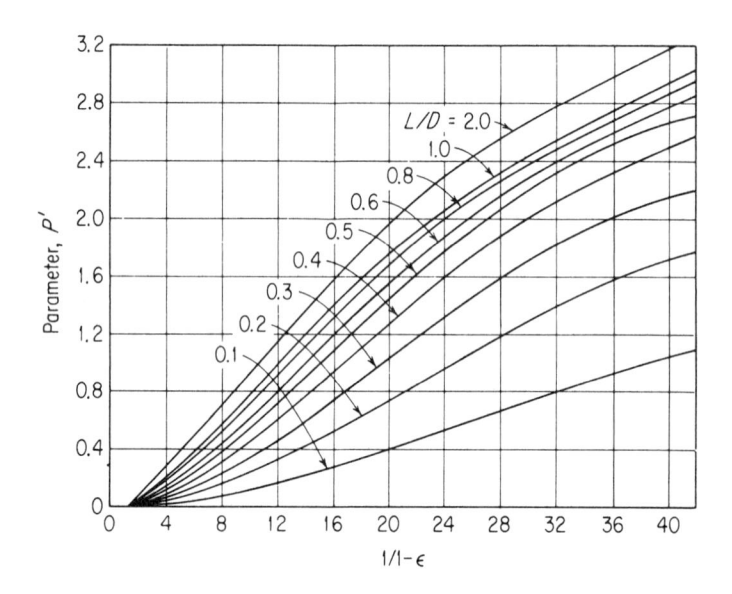

Fig. 6-7. Bearing parameter, P', vs. eccentricity ratio, $1/(1 - \epsilon)$—journal bearings.

FLUID MECHANICS

What torque will be required to rotate this shaft at 500 r/min?

a) From Fig. 6-6,[3] below 600 r/min, the recommended value of "operating diametral clearance" is $c_d = 0.088$ mm.
b) The resultant "modulus" is: $m = c_d/d = 0.088/50 = 0.00176$ mm/mm.
c) The "bearing pressure parameter" is: $P' = 1000\ m^2 p_b/\mu N$ in which

p_b = average force on projected area of bearing   kPa
μ = dynamic (absolute) viscosity   Pa·s
N = revolutions per minute   r/min

Assume heavy line shafting as shown in Table 6-4,[3] then $p_b = 1050$ kPa. The viscosity $\mu = 34 \times 10^{-6}(880) = 0.030$ Pa·s. Thus: $P' = 1000(0.00176)^2 1050/0.030(500) = 0.22$.

d) Recognizing $L/d = 75/50 = 1.5$, and applying P' to Fig. 6-7[3] to obtain a value for the "eccentricity ratio" ϵ, it is observed that: $1/(1 - \epsilon) = 4.0$.
e) Entering Fig. 6-8[3] with this value, the "torque parameter" is seen to be $T' = 1.25$.
f) The torque required to rotate the shaft under these conditions may then be obtained by the expression:

$$T = T' r^2 \mu N/m$$

$$T = 1.25(0.025)^2(0.030)500/0.00176$$

$$= 6.66 \text{ N·m/rad·m}$$

but since the bearing is 0.075 m in length, the actual torque is:

$$T = 6.66(0.075) = 0.50 \text{ N·m/rad}$$

Note that this analysis includes a moderate bearing pressure while Example 6-23 was assumed as a "no-load" condition. Thus the results compare favorably.

The power required to overcome friction in the bearing may now be easily determined as:

$$P = 2\pi nT = 2\pi(500/60)0.50 = 26.2 \text{ W}$$

$$\frac{\text{rad}}{\text{r}} \cdot \frac{\text{r}}{\text{s}} \cdot \frac{\text{N·m}}{\text{rad}} = \text{N·m/s} = \text{J/s} = \text{W}$$

[3] Ibid.

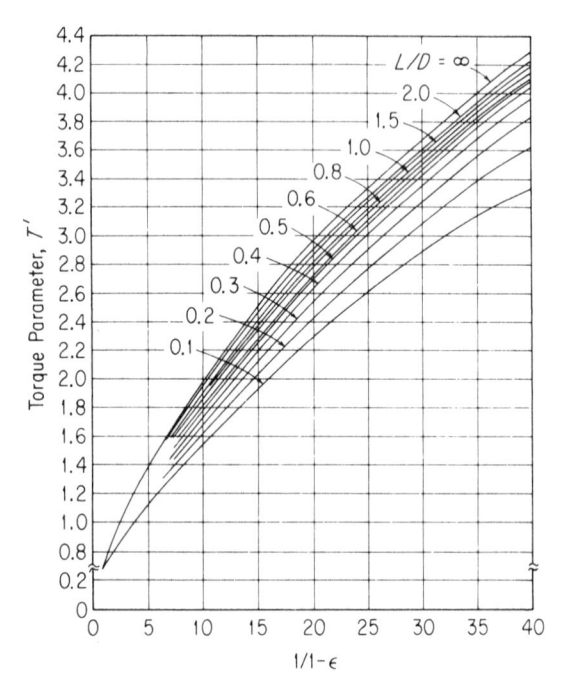

Fig. 6-8. Torque parameter, T', vs. eccentricity ratio, $1/(1 - \epsilon)$—journal bearings.

6.25 Pump and Turbine Characteristics

The basis of the theory of pumps and turbines is Euler's Equation which may be written:

$$T = Q\rho(u_1 r_1 - u_2 r_2)$$

where T is the torque exerted or experienced by the rotor, Q is the volumetric flow rate through the device, ρ is the density of the fluid; u_1 and u_2 are the tangential components of the fluid velocity at the entry and exit radii r_1 and r_2, respectively. In terms of units,

$$\frac{m^3}{s} \cdot \frac{kg}{m^3} \cdot \frac{m}{s} \cdot m = \frac{kg \cdot m}{s^2} \cdot m \cdot \frac{m}{m} = N \cdot m/rad$$

Hence, the principles as expressed in Euler's Equation may be applied directly in SI units as shown in the following examples. Note that an m/m ratio is retained in the equation to provide for the correct units of torque in N·m/rad.

Example 6-25

A centrifugal pump delivers water at 0.2 m³/s against a head of 4.5 m. Assuming an overall efficiency of 85 per cent calculate the power consumed. If the water enters the pump with zero tangential velocity and leaves it at an average radius of 280 mm, estimate a reasonable diameter for the delivery connection. The useful power is given by:

(Eq. 6.26) $$P = Q\rho gh$$
$$P = 0.2(1000)(9.8)4.5 = 8830 \text{ W}$$

$$\text{power consumed } \frac{8830}{0.85} = 10.4 \text{ kW}$$

$$\frac{m^3}{s} \cdot \frac{kg}{m^3} \cdot \frac{m}{s^2} \cdot m = \frac{N \cdot m}{s} = \frac{J}{s} = W$$

Referring to Equation (3.22) the power corresponds to a torque T multiplied by an angular velocity ω, so that in this case $T = 10400/\omega$ and since $u_1 = 0$:

$$P = T\omega;$$

$$T = \frac{10400}{\omega} = (0.2)(1000)(u_2)0.28 = 56u_2 \quad \text{N·m/rad}$$

Suppose the radius of the rotor is 260 mm; then the maximum tangential velocity component u_2 of the water will be 0.26ω m/s. From Euler's Equation again:

$$\frac{10400}{\omega} = (56)0.26\omega$$

whence

$$\omega = \left[\frac{10\,400}{(56)0.26}\right]^{1/2} = 26.7 \text{ rad/s}$$

and

$$u_2 = 0.26\omega = 6.9 \text{ m/s}$$

For a delivery of 0.2 m³/s the area of the delivery connection should be:

$$0.2/6.9 = 0.029 \text{ m}^2 \qquad \frac{m^3}{s} \cdot \frac{s}{m} = m^2$$

which corresponds to a diameter of about 190 mm.

Table 6-5. Variations in Fluid Flow Formulas According to Unit Systems (illustrative)

	Customary Units	SI Units (a)	SI Units (b)
General Equation	L = ft h = ft Q = cfs A = ft^2 $\bar{V}$ = ft/s R = ft	L = m h = m Q = m^3/s A = m^2 $\bar{V}$ = m/s R = m	L = mm h = mm Q = L/s A = mm^2 $\bar{V}$ = m/s R = mm
Standard Orifice: $Q = C_d A \sqrt{2g}\, h^{1/2}$	$Q = 8.02 C_d A h^{1/2}$	$Q = 4.43 C_d A h^{1/2}$	$Q = 140.1(10^{-6}) C_d A h^{1/2}$
	C_d varies from 0.59 to 0.64 depending on head/size ratio and orifice shape.		
Rectangular Weir: (suppressed; sharp crest) $Q = C_d (2/3)\sqrt{2g}\, L h^{3/2}$	$Q = 3.32 L h^{3/2}$ assuming $C_d = 0.62$	$Q = 1.83 L h^{3/2}$ $C_d = 0.62$	$Q = 57.9(10^{-6}) L h^{3/2}$ $C_d = 0.62$

V-Notch Weir:

$$Q = C_d \frac{8}{15} \tan \alpha \sqrt{2g} \; h^{5/2} \qquad Q = 2.5h^{5/2} \qquad Q = 1.38h^{5/2} \qquad Q = 43.7(10^{-6})h^{5/2}$$

for $2\alpha = 90°$ and $C_d = 0.585$ in all cases

Open Channel:
(Manning)

$$\bar{V} = \frac{k}{n} R^{2/3} S^{1/2} \qquad \bar{V} = \frac{1.486}{n} R^{2/3} S^{1/2} \qquad \bar{V} = \frac{1}{n} R^{2/3} S^{1/2} \qquad *\bar{V} = \frac{0.01}{n} R^{2/3} S^{1/2}$$

*for $\bar{V}$ in m/s where R is in mm

Conduits:
(Hazen-Williams)

$$\bar{V} = kCR^{0.63} S^{0.54} \qquad \bar{V} = 1.318 CR^{0.63} S^{0.54} \qquad \bar{V} = 0.849 CR^{0.63} S^{0.54} \qquad *\bar{V} = 1.094(10^{-2}) CR^{0.63} S^{0.54}$$

6.26 Orifices, Weirs, Channels and Conduits

All principles and procedures for determining flow through orifices, over weirs, in channels and in conduits apply in SI in the same manner as in the customary measurement system but all of the so-called "constants" change in accordance with the units and multiples of units used as is indicated in the comparison of formulas given in Table 6-5.

In the instances of orifices and weirs the change in the numerical value of the acceleration of gravity represents the principal difference. As is clearly reflected in the case of the standard orifice formula, the value of $\sqrt{2g}$ which in foot units is 8.02, becomes $[2(9.8)]^{1/2} = 4.43$ when head is expressed in meters, and $[2(9810)]^{1/2} = 140.1$ when head is expressed in millimeters.

When in the millimeter form of the equation it is desired to express the value of Q in liters per second (L/s) it is also necessary to multiply by 10^{-6} which represents the ratio between a liter and cubic millimeters. Occasionally the multiplier used is 10^{-5} but in that case the decimal point is shifted accordingly in the prefatory constant. Emphasis on use of 10^{-6} is generally recommended.

In the case of weirs these same situations pertain but there are additional numerical considerations dependent solely on the geometry of the weir. These geometric constants, however, have the same value in all measuring systems.

A somewhat different circumstance is faced in the widely used Manning Equation for velocity of flow in open channels. The desired dimensions for $\bar{V}$ are L/t. When the dimensions of R and S are introduced to the powers indicated, inherently the factor n must take on the dimensions $t/L^{0.33}$. Fortunately the unit of time, the second (s), is the same in all measurement systems. But the unit of length to be used will require consideration in the writing of the formula. Manning's data was in metric units. Accordingly, the familiar value of the roughness factor for a concrete pipe, with its proper units is: $n = 0.013$ s/m$^{0.33}$. To use the widely published values of Manning's n in the customary system required the introduction of the factor 1.486 which is literally the ratio of feet per meter to the one-third power, i.e. $(3.281)^{0.33} = 1.486$. For those who have been using the Manning expression in the customary unit form, changeover to SI will be accompanied by the *dropping* of this conversion factor of 1.486.

Also when the Manning formula is used with R in millimeters and the desired result is $\bar{V}$ in m/s, a correction factor of 10^{-2} must be included in the numerator to compensate for the $(1000)^{0.67}$ ratio introduced through the R term.

Similar situations will be found in the use of other empirical formulas. Whenever dimensional analysis indicates factors with inherent units, a change in the numerical value should be anticipated and checked out. For further discussion, see Chapter 10.

7
Thermodynamics and Heat Transfer

7.0 Introduction

Assuming that the reader has a good understanding of engineering principles, no attempt will be made here to expound on the very subtle and far-reaching subjects of thermodynamics and heat transfer. Instead, the emphasis will be on a few examples of the very favorable impact of SI on representative calculations. Regarded solely from the standpoint of units, this subject emphasizes principally two additional SI aspects, namely, temperature and heat.

A listing of Quantities and Preferred SI Units as used in Thermodynamics and Heat Transfer is given in Table 7-1 and some Frequently Used Formulas are given in Table 7-2.

7.1 Temperature

By definition of the kelvin (K) as a Base Unit, the temperature scale in SI is the absolute scale in centigrade intervals. The terminology has changed slightly; the degree is known as the "kelvin," with the symbol K, whether it refers to an actual temperature or to a temperature difference. Thus the freezing point and boiling point of water under ordinary conditions are 273 K, and 373 K, respectively, and the difference between them is 100 K. Note especially that the word "degree" and the symbol ° are *not* used with the kelvin. The Celsius scale in which the freezing point of water is the zero reference, is not directly a part of SI but is used for stating other than thermodynamic temperature, subject to conversion when thermodynamic phenomena are involved. The special name "degree Celsius" (symbol °C) is given to a unit of this scale; it is equal to a kelvin.

7.2 Pressure

Another key aspect of thermodynamics to be kept in mind is that while pressure generally is stated as "gage" pressure, the proper

Ch. 7 THERMODYNAMICS AND HEAT TRANSFER

Table 7-1. Quantities and SI Units Frequently Used in Thermodynamics and Heat Transfer[**]
(References: ANSI Y 10.4-1957; ISO R31 Part IV-1960)

A = area	m²
c = mass concentration	kg/m³
c = unit conductance k/L	W/(m²·K)
c = specific heat capacity, general (liquids and solids)	J/(kg·K)
c_p, c_v = specific heat capacity, constant p, constant v (gases)	J/(kg·K)
C = heat capacity*	J/K
C = molar concentration	mol/m³
D = mass diffusion coefficient, {(kg/s)(m)(m³/kg)(m⁻²)} (mass diffusivity: $D = \dot{m}L/\rho A$)	m²/s
D_T = thermal diffusion coefficient {(J/s)(m⁻²)(m/K) (m³/kg)(kg·K/J)} (for mass diffusion induced by temperature gradient: $D_T = k/\rho c$)	m²/s
e = specific energy*	J/kg
E = energy*	J
G = mass velocity, mass current density {(kg/m³)(m/s)} ($G = \rho \bar{V}$) or ($G = \dot{m}/A$)	kg/(m²·s)
G = irradiation (incident radiant flux per unit area, per unit time)	W/m²
h = surface heat transfer coefficient; film coefficient (rate of heat transfer per unit area per unit temperature difference across a boundary surface) {(J/s)(m⁻²)(K⁻¹)}	W/(m²·K)
h_{fg} = specific enthalpy of vaporization	J/kg
h = specific enthalpy $(u + pv)$	J/kg
h_m = molar enthalpy	J/mol
H = enthalpy	J
k = thermal conductivity {(J/s)(m⁻²)(m/K)}	W/(m·K)
K = thermal conductance $K = kA/L$	W/K
L = length, distance	m
m = mass	kg
$\dot{m}$ = mass flow rate ($\dot{m} = \rho \bar{V}$) {(kg/m³)(m³/s)}	kg/s
M = molecular mass	kg/mol
p = absolute pressure; alternate: $p(abs)$	Pa
p_g = gage pressure; alternate $p(gage)$	Pa
p_o = atmospheric pressure	Pa

* Form may be denoted by various subscripts and superscripts.
** For complete interpretation of this table, note the following in terms of Base Units:

$$J/kg = m^2/s^2 \qquad J = N \cdot m = kg \cdot m^2/s^2$$
$$J/K = kg \cdot m^2/(s^2 \cdot K) \qquad J/s = W = kg \cdot m^2/s^3$$
$$J/(kg \cdot K) = m^2/(s^2 \cdot K) \qquad Pa = N/m^2 = kg/(m \cdot s^2)$$

Table 7-1. Quantities and SI Units Frequently Used in Thermodynamics and Heat Transfer** *(Continued)*

P = power (energy, heat, or work rate)	J/s = W
q = heat transfer rate; alternate to P ($q = \Delta T/R$)	J/s = W
q'' = heat flux density (heat transfer rate per unit area)	J/(s·m²) = W/m²
Q = heat transfer across the boundary of a system	J
R = thermal resistance ($R = \Delta T/q$); $R = L/Ak$ {(m/m²)(m·K/W)}	K/W
R_u = universal gas constant	J/(mol·K)
R = specific gas constant ($R = R_u/M$)	J/(kg·K)
s = specific entropy	J/(kg·K)
S = entropy	J/K
t = time	s
t = temperature	°C
T = thermodynamic, or absolute, temperature	K
ΔT = interval of thermodynamic temperature	K
u = specific internal energy	J/kg
u_m = molar internal energy	J/mol
U = internal energy of a system	J
U = overall heat transfer coefficient ($U = q/A\cdot\Delta T$) (overall unit conductance; overall unit transmittance)	W/(m²·K)
v = specific volume	m³/kg
V = volume	m³
$\dot{V}$ = volumetric rate of flow	m³/s
V_m = molar volume	m³/mol
$\vec{V}$ = velocity	m/s
$\dot{W}$ = work rate (occasional alternate for power: P)	J/s
W = work	J
x = quality; mass fraction of a vapor in a two-phase mixture; dryness fraction	(ratio)
z = elevation above a datum	m

Greek Symbols

γ = heat capacity ratio ($\gamma = c_p/c_v$)	(ratio)
ϵ = emittance, emissivity (radiation emittance)	(ratio)
σ = Stefan-Boltzmann constant ($\sigma = 5.670 \times 10^{-8}$)	W/(m²·K⁴)
ρ = density	kg/m³

Subscripts

c = critical state; f = saturated liquid; g = saturated vapor; fg = change of phase at constant pressure; i = saturated solid; s = saturation state; m = molar; o = overall; g = gage.

Table 7-2. Frequently Used Formulas in Thermodynamics and Heat Transfer

Number	Formula
(7.1)	$q = \dot{m} c\, \Delta T$
(7.2)	$Q - W = m \cdot \Delta(u + \frac{1}{2}\bar{V}^2 + pv)$ (when $\Delta E_p = 0$)
(7.3)	$h = u + pv$
(7.4a)	$Q - W = m\{(h_2 - h_1) + \frac{1}{2}(\bar{V}_2^2 - \bar{V}_1^2) + g(z_2 - z_1)\}$ (steady-state, steady flow)
(7.4b)	$q - P = \dot{m}\{(h_2 - h_1) + \frac{1}{2}(\bar{V}_2^2 - \bar{V}_1^2) + g(z_2 - z_1)\}$ (steady-state, steady flow)
(7.5)	$v_1 = v_f + x v_{fg}$
(7.6)	$u_1 = u_f + x u_{fg}$
(7.7)	$h_1 = h_f + x h_{fg}$
(7.8)	$s_1 = s_f + x s_{fg}$
(7.9)	$S_2 - S_1 = \int_1^2 \left(\dfrac{dQ}{T}\right)$, or, $dS = \left(\dfrac{dQ}{T}\right)$ (reversible process)
(7.10)	$T ds = du + p dv$
(7.11)	$T ds = dh - v dp$
(7.12)	$s_2 - s_1 = c_p \ln(T_2/T_1)$ (constant pressure process)
(7.13)	$s_2 - s_1 = c_v \ln(T_2/T_1)$ (constant volume process)
(7.14)	$c_v = du/dT\vert_v$
(7.15)	$c_p = dh/dT\vert_p$
(7.16)	$pv = RT$, or, $pV = mRT$
(7.17)	$c_p - c_v = R$ (ideal gas)
(7.18)	$pv^\gamma = $ constant (isentropic process, ideal gas)
(7.19)	$\gamma = c_p/c_v$
(7.20)	$W/m = (p_1 v_1 - p_2 v_2)\gamma/(\gamma - 1)$ (adiabatic process)
(7.21)	$q = k \cdot A \cdot \Delta T/\Delta x$
(7.22)	$q = h \cdot A (T_1 - T_2)$
(7.23)	$q = F_G F_\epsilon \sigma A (T_1^4 - T_2^4)$
(7.24)	$q = U \cdot A \cdot \Delta T$
(7.25)	$1/U = R_T = R_1 + R_2 + R_3$, etc.
(7.26)	$R_T = \dfrac{1}{h_1} + \dfrac{L_2}{k_2} + \dfrac{L_3}{k_3} + \dfrac{L_4}{k_4} + \dfrac{1}{h_2}$, etc.

pressure p to be used in thermodynamics computations is the *absolute* pressure. For this purpose it is well to note that normal atmospheric pressure, as found in most engineering operations, is about 0.1 MPa, or 100 kPa. Actually, the standard atmosphere is 101.325 kPa. Atmospheric pressure must be added algebraically to any given gage pressure to obtain absolute pressure. For a further discussion of "pressure" see Section 6.2.

7.3 Heat

Heat is energy transferred between a body and its surroundings because of a temperature difference. Thus heat is energy in transition. The unit used for heat is identical with the unit for work because of the equivalence of heat and work established by the First Law of Thermodynamics, generally stated in the following manner: "If a system is taken through a cycle of changes (processes) so that it returns to the same state or conditions from which it started, the sum of the heat and work effects will be zero."

The sole SI unit for the measurement of all forms of energy is the joule (J). Since the joule is the same as the newton-meter, statements such as "the mechanical equivalent of one British thermal unit is 778 ft-lbf" become trivial; in SI the mechanical equivalent of one joule is one newton-meter. The joule is also the watt-second, so that mechanical, thermal, and electrical energy are all expressed in terms of the same unit in SI. This leads to a considerable reduction of the arithmetic in many thermodynamics and heat transfer calculations.

7.4 Derived Units

As would be anticipated, physical quantities in thermodynamics and heat transfer have units based principally on: energy, temperature or temperature differentials, mass, time, pressure, and volume in accordance with the definition of each particular quantity. The following are a few illustrations:

C = heat capacity of a body, i.e., the quantity of heat required to raise its temperature by one degree, is expressed in joules per kelvin   J/K

c = specific heat capacity, i.e., the heat capacity per unit mass is expressed in joules per kilogram-kelvin   J/(kg·K)

q = the heat transfer rate is expressed in joules per second, which in terms of SI units yields the watt   J/s = W

q'' = the heat transfer rate per unit area, sometimes termed heat flux density, is expressed in joules per meter-squared-second, which in terms of SI units yields watts per square meter   J/(m²·s) = W/m²

h = the heat transfer rate per unit surface area produced by unit temperature difference (sometimes termed "film coefficient") is   W/(m²·K)

As might be expected, quite different phenomena may have, in

Ch. 7   THERMODYNAMICS AND HEAT TRANSFER   195

ultimate form, the same final read-out of units. Thus, before coming to any technical conclusions about such a coincidence, full consideration should be given to the exact definition of each phenomenon. A complete text on the principles of thermodynamics or on heat transfer should be consulted for each purpose. Also it should be noted that there are a number of letter symbols which by convention are sometimes used to express one "quantity" and at other times to express another quantity. Accordingly, each symbol and each formula must always be interpreted in the context in which it is used or stated. Examples in SI terms which follow will demonstrate these observations.

Example 7-1

Energy at the rate of 5.0 kW is being used to heat a liquid flowing through a pipe at 2.0 kg/s. If the specific heat capacity of the liquid is 2100 J/(kg·K), what will be the resulting rise in temperature?

The applicable expression in thermodynamic notation is:

(Eq. 7.1) $q = \dot{m} c \Delta T$ $\dfrac{kg}{s} \cdot \dfrac{J}{kg \cdot K} \cdot K = \dfrac{J}{s} = W$

Since in one second 2.0 kg of liquid will receive 5000 joules of heat, the temperature rise accordingly is:

$\Delta T = q/\dot{m}c$ $\dfrac{5000}{2.0(2100)} = 1.2 \text{ K}$ $W \cdot \dfrac{s}{kg} \cdot \dfrac{kg \cdot K}{J} = K$

as the watt-second is equivalent to the joule.

Example 7-2

A liquid flowing at 5.0 kg/s through a horizontal pipe suffers an overall loss of head of 15.0 m of fluid through pipe friction. Assuming that the pipe is perfectly insulated, what will be the resulting rise in temperature? Take the density of the liquid as $\rho = 1300$ kg/m³ and its specific heat capacity, c, as 960 J/(kg·K).

Considering first the fluid flow: $\dot{V} = \dot{m}/\rho$ whence the volumetric flow rate is:

$\dot{V} = 5.0/1300 = 0.00385$ m³/s $\dfrac{kg}{s} \cdot \dfrac{m^3}{kg} = \dfrac{m^3}{s}$

The pressure drop corresponding to a loss of head of 15.0 m of fluid is

196   THERMODYNAMICS AND HEAT TRANSFER   Ch. 7

found, using Equation (6.1):

$$\Delta p = \rho g h,$$

whence

$$\Delta p = 1300(9.8)15 = 0.191 \text{ MPa} \qquad \frac{\text{kg}}{\text{m}^3} \cdot \frac{\text{m}}{\text{s}^2} \cdot \text{m} = \frac{\text{N}}{\text{m}^2} = \text{Pa}$$

The transfer of energy per second, q, may be stated as:

$$q = \dot{V}\Delta p$$

$$q = 0.191 \times 10^6 (0.00385)$$

$$= 735 \text{ W} \qquad \frac{\text{N}}{\text{m}^2} \cdot \frac{\text{m}^3}{\text{s}} = \frac{\text{N} \cdot \text{m}}{\text{s}} = \frac{\text{J}}{\text{s}} = \text{W}$$

Applying Equation (7.1):

$$\Delta T = \frac{q}{\dot{m}(c)} = \frac{735}{5.0(960)} = 0.153 \text{ K} \qquad \text{W} \cdot \frac{\text{s}}{\text{kg}} \cdot \frac{\text{kg} \cdot \text{K}}{\text{J}} = \text{K}$$

Note the facility with which this kind of calculation is performed: the arithmetic is reduced to an absolute minimum, largely because of the use of the same unit for both work and heat. The simplification introduced by using SI units is particularly striking in thermodynamics, and the increased clarity of the calculations is also noteworthy.

7.5 Steady-Flow Energy Equation

The preceding examples are extremely simple applications of a basic thermodynamic equation known as the "steady flow energy equation." In words, this states that for a fluid under steady flow conditions, for a given interval of time, the heat added, Q, minus the work done, W, is a function of the increase in energy, e, plus the increase in the work, pv. In general form, an equation may be written:

$$Q - W = m \cdot \Delta(e + pv),$$

where m is the mass of fluid flowing in the time interval. Further, it can be shown that the fluid energy (e) may be comprised of a number of components, including: kinetic energy, potential energy, internal energy, electrical energy, and so on. But for a wide range of situations the significant terms are the kinetic and internal energies.

The equation may then be written:

(Eq. 7.2) $Q - W = m \cdot \Delta(u + \tfrac{1}{2}\bar{V}^2 + pv),$ kg(J/kg) = J

Ch. 7   THERMODYNAMICS AND HEAT TRANSFER   197

where u is the specific internal energy and $\bar{V}$ the average velocity. Further, the sum of the terms u and pv is a property called "enthalpy" which is denoted by the symbol, h.

(Eq. 7.3) $\qquad h = u + pv \qquad$ J/kg

When using SI units all of the principal parts of Equation (7.2) are expressed in terms of joules. On the left-hand side, as previously observed, both heat and work will be expressed in joules. On the right-hand side it is essential to observe that since m is in kg, the following relationships result:

u = specific internal energy $\qquad$ J/kg; $\quad$ hence $m\Delta u \quad$ J
$\frac{1}{2}\bar{V}^2$ = specific kinetic energy $\qquad$ m²/s²; $\quad$ thus $m\Delta\frac{1}{2}\bar{V}^2 \quad$ J
p = pressure $\qquad$ $Pa = \dfrac{N}{m^2} = kg/(m \cdot s^2)$;
v = specific volume $\qquad$ m³/kg;

and therefore $m \cdot \Delta(pv)$ is: $\qquad kg \cdot \dfrac{kg}{m \cdot s^2} \cdot \dfrac{m^3}{kg} = \dfrac{kg \cdot m^2}{s^2} = N \cdot m = J$

The enthalpy, h, is also "specific," having the unit J/kg.

The steady-flow energy equation may be found and used in any of several forms. The most general form is Equation (7.4a). This is frequently rewritten in terms of unit time, in which case q and P are the heat transfer rate and the rate of external work, respectively, and $\dot{m}$ is the mass rate of flow, as is shown in Equation (7.4b).

(Eq. 7.4a) $\quad Q - W = m\{(h_2 - h_1) + \frac{1}{2}(\bar{V}_2^2 - \bar{V}_1^2) + g(z_2 - z_1)\} \quad$ J

(Eq. 7.4b) $\quad q - P = \dot{m}\{(h_2 - h_1) + \frac{1}{2}(\bar{V}_2^2 - \bar{V}_1^2) + g(z_2 - z_1)\} \quad \dfrac{J}{s} = W$

To assure the correct application of each of the above equations an analysis of the applicable units is advisable.

For Equation (7.4a), remembering that m²/s² = J/kg, the unit analysis is:

$$J - J = kg\left\{\left(\dfrac{J}{kg}\right) + \left(\dfrac{m^2}{s^2}\right) + \left(\dfrac{m}{s^2} \cdot m\right)\right\} = J$$

For Equation (7.4b), it will be noted that q is in J/s; also that the rate of work (occasionally shown as $\dot{W}$) is here shown directly as P indicating "power," which is in J/s. With $\dot{m}$ being stated in kg/s, the result of the unit analysis is, of course, in watts (W).

Further it may be noted that if m is removed from Equation (7.4a) the result becomes J/kg.

198   THERMODYNAMICS AND HEAT TRANSFER   Ch. 7

Example 7-3

The previous example can be reconsidered in the light of the steady-flow energy equation. The various quantities have the following values: $Q = 0$, $W = 0$, since no heat is added or work done; $\dot{m} = 5.0$ kg/s; Δu the increase in internal energy, equals the specific heat capacity, c, 960 J/(kg·K) multiplied by the temperature rise, ΔT.

In view of the fact that no reference is made to any change in velocity, the component $\Delta(\bar{V}^2/2)$ is omitted. Thus, since the specific volume is the reciprocal of the density and since the increase in pressure is negative:

$$\Delta(pv) = \Delta p/\rho = 0.191 \times 10^6/1300 \qquad \text{Pa} \cdot \frac{m^3}{kg} = \frac{N}{m^2} \cdot \frac{m^3}{kg} = \frac{J}{kg}$$

Applying Equation (7.2):

$$\Delta(u + pv) = 0$$

$$5.0[(960 \cdot \Delta T) + (-0.191 \times 10^6/1300)] = 0$$

$$\Delta T = 0.153 \text{ K}$$

$$\frac{kg}{s}\left[\left(\frac{J}{kg \cdot K} \cdot K\right) + \left(\frac{N}{m^2} \cdot \frac{m^3}{kg}\right)\right] = \frac{J}{s} + \frac{N \cdot m}{s} = \frac{J}{s} = W$$

But thermodynamics is much more concerned with gases than with liquids. For most situations dealing with gases it is necessary to consider the dependence of specific heat on temperature and in many cases on pressure.

Example 7-4

A fluid flows through an engine at the rate of $\dot{m} = 0.6$ kg/s. Other conditions are as follows:

	At Entry	At Exit
pressure (absolute)	$p_1 = 3.0$ MPa	$p_2 = 0.5$ MPa
specific volume	$v_1 = 0.0993$ m³/kg	$v_2 = 0.4252$ m³/kg
specific internal energy	$u_1 = 2933$ kJ/kg	$u_2 = 2644$ kJ/kg

Assume that the differential between entrance and exit velocities is negligible. Heat is supplied to the engine at the rate of 25 kW. What is the indicated power output?

Applying Equations (7.3) and (7.4a) in adjusted form:

$$q - P = \dot{m}\{(u_2 - u_1) + (p_2 v_2 - p_1 v_1)\}$$

Ch. 7 THERMODYNAMICS AND HEAT TRANSFER 199

The energy terms on the right-hand side are:

$$(u_2 - u_1) = (2644 - 2933) = -289 \quad \text{kJ/kg}$$

$$(p_2 v_2 - p_1 v_1) = 0.5 \times 10^6 (0.4252) - 3.0 \times 10^6 (0.0993)$$

$$= -0.0853 \times 10^6 \quad \text{J/kg}$$

whence:

$$25\,000 - P = 0.6(-289 \times 10^3 - 0.0853 \times 10^6) \quad \frac{\text{kg}}{\text{s}} \cdot \frac{\text{J}}{\text{kg}} = \frac{\text{J}}{\text{s}} = \text{W}$$

$$P = 249.6 \text{ kW}$$

The chief purpose of including this somewhat ideal example in this form is to demonstrate again the way in which SI units such as the watt, joule, and pascal can be linked in a single equation without the need for any conversion factors. In contrast, in English units, if q were given in watts, P in horsepower, $\dot{m}$ in lb/min, u in Btu/lb, p in lbf/in.^{2} and v in ft^3/lb, a whole series of conversion factors would have to be introduced to achieve a common basis.

Of course, a further simplification might be to use enthalpies directly, if available, from an appropriate table or chart, many of which are now becoming available in SI units for most of the more common working substances.

In this example, only the data on pressures and temperatures would normally be available through actual measurement. Other reference data required to complete the components of the energy equation will involve the known physical properties of the fluid. Of the reference sources, steam tables are among the most important, and afford an excellent example for consideration from the standpoint of units.

7.6 Use of Steam Tables

Steam tables present data on specific volume, internal energy, enthalpy, and entropy as functions of absolute pressure and temperature. When SI units are used, temperature is customarily in degrees Celsius and the absolute pressure in pascals, though some tables have used the bar, which is 10^5 Pa. Specific volume is in m^3/kg; specific internal energy and enthalpy in kJ/kg; and entropy in kJ/(kg·K). Otherwise, the tables are laid out exactly as for any other system of units, and are used in precisely the same way. Two principal warnings in using the steam tables, however, are to observe that pressures are to be stated in absolute terms (keeping in mind that atmospheric pressure is approxi-

Table 7-3. Some Properties of Water and Steam

800 kPa		$t_s = 170.41°C$	
	v_g 240.3	h_g 2768.0	s_g 6.6594
t °C	v *10^{-3} m³/kg	h kJ/kg	s kJ/(kg·K)
0	0.9998	0.8	0.0001
10	0.9999	42.8	0.1511
20	1.0013	84.6	0.2962
30	1.0039	126.3	0.4361
40	1.0075	168.0	0.5741
50	1.0118	209.8	0.7027
60	1.0169	251.6	0.8300
70	1.0226	293.4	0.9538
80	1.0289	335.3	1.0742
90	1.0359	377.3	1.1914
100	1.0435	419.4	1.3058
110	1.0516	461.6	1.4174
120	1.0604	504.0	1.5265
130	1.0699	546.5	1.6333
140	1.0800	589.2	1.7379
150	1.0908	632.1	1.8405
160	1.1023	675.3	1.9414
170	1.1147	718.8	2.0407
**			
180	247.3	2794.2	6.7178
190	254.3	2818.7	6.7714
200	261.0	2841.8	6.8207

* Actual $v_g = 0.2403$, and $v_f = 0.0011147$ m³/kg.
** Flashpoint of water.

mately 100 kPa) and to watch for the presence of decimal multiplying factors or prefixes. For example, in some tables the specific volumes tabulated are 100, or 1000 times the true values, to avoid taking up space with zeroes. Usually, values of internal energy, enthalpy, and entropy will be given in terms of the kilojoule, kJ, rather than the joule (see Table 7-3).

Where reference tables in customary units are the only ones presently available it will be well to observe that:

$$2.326 \text{ kJ/kg} = 1.0 \text{ Btu/lb}$$

$$4.1868 \text{ kJ/(kg·K)} = 1.0 \text{ Btu/(lb·°R)}$$

Ch. 7   THERMODYNAMICS AND HEAT TRANSFER

Example 7-5

A boiler is generating steam at a rate of $\dot{m} = 2.60$ kg/s at a pressure of 900 kPa (gage) and a temperature of 180°C. The fuel being used has a heat of combustion of 29 300 MJ/m^3 and it is burned at 75 percent efficiency. The atmospheric pressure may be taken as 100 kPa. The temperature of the feed water is 93°C. What is the required rate of consumption of fuel oil?

$$p_2 = 900 + 100 = 1000 \text{ kPa (abs)}$$

From typical steam tables as shown in Table 7-3 and Table 7-5:

$h_1 =$ enthalpy of water at 93°C $= 390$ kJ/kg

$h_2 =$ enthalpy of steam at 1000 kPa (abs) $= 2780$ kJ/kg

Applying Equation (7.4b):

$$q - P = \dot{m}[(h_2 - h_1) + \tfrac{1}{2}(\bar{V}_2^2 - \bar{V}_1^2) + g(z_2 - z_1)]$$

and observing that $\bar{V}_2 - \bar{V}_1 = 0$; $z_2 - z_1 = 0$; $P = 0$,

$$q/\dot{m} = (h_2 - h_1) = 2780 - 390 = 2390 \text{ kJ/kg}$$

$$q = 2.60(2390) = 6214 \text{ kJ/s}$$

$$\text{Fuel oil required} = \frac{6214 \times 10^3}{0.75(29\,300 \times 10^6)} = 0.283 \text{ L/s} = 1.019 \text{ m}^3/\text{h}.$$

7.7 Quality or Dryness Fraction

Thermodynamic problems frequently involve a mixture of saturated liquid and of saturated vapor. Such conditions may be described by a "quality" or "dryness fraction" for which the common symbol is x. A unit mass of wet vapor can be analyzed as a simple mixture of saturated liquid and saturated vapor since the specific properties of each are retained in such a mixture. Equations of the form of (7.5), (7.6), (7.7) and (7.8) are frequently used in applying reference data such as that given for illustrative purposes in Tables 7-3, 7-4, and 7-5. Notation is as given on the tables. The basis of this form of analysis is as indicated in Fig. 7-1. Designating: $x =$ quality, mass fraction of vapor in a two-phase system, the resultant specific values for the conditions denoted by point

Table 7-4. Steam Table: Saturation—(a) by Temperature

Temp.	Press.	Specific Volume		Specific Internal Energy		Specific Enthalpy			Specific Entropy		
t	p_s	v_f	v_g	u_f	u_g	h_f	h_{fg}	h_g	s_f	s_{fg}	s_g
°C	kPa	10^{-3} m³/kg		kJ/kg		kJ/kg			kJ/(kg·K)		
10	1.23	1.00	106 000	42.0	2390	42.0	2480	2520	0.151	8.75	8.90
20	2.34	1.00	57 800	83.9	2400	83.9	2450	2540	0.296	8.37	8.67
30	4.24	1.00	32 900	126	2420	126	2560	2560	0.436	8.02	8.45
50	12.3	1.01	12 000	209	2440	209	2590	2590	0.704	7.37	8.07
100	101	1.04	1 670	419	2510	419	2680	1680	1.31	6.05	7.36
200	1550	1.15	127	850	2600	852	2790	2790	2.33	4.10	6.43

Subscript notation: s = at saturation; f = saturated liquid; g = saturated vapor; fg = vaporization.

Table 7-5. Steam Table: Saturation—(b) by Pressure

Press.	Temp.	Specific Volume		Specific Internal Energy		Specific Enthalpy			Specific Entropy		
p	t_s	v_f	v_g	u_f	u_g	h_f	h_{fg}	h_g	s_f	s_{fg}	s_g
kPa	°C	10^{-3} m³/kg		kJ/kg		kJ/kg			kJ/(kg·K)		
1	7.0	1.01	129 000	29.3	2380	29.3	2480	2510	0.106	8.86	8.97
2	17.5	1.00	67 000	73.4	2400	73.4	2460	2530	0.260	8.46	8.72
10	45.8	1.01	14 700	192	2440	192	2390	2580	0.649	7.50	8.15
20	60.1	1.02	7 650	251	2460	251	2360	2610	0.832	7.07	7.91
100	99.6	1.04	1 690	418	2510	418	2260	2670	1.30	6.06	7.36
1000	180	1.13	194	762	2584	763	2020	2780	2.14	4.45	6.59

Subscript notation: s = at saturation; f = saturated liquid; g = saturated vapor; fg = vaporization.

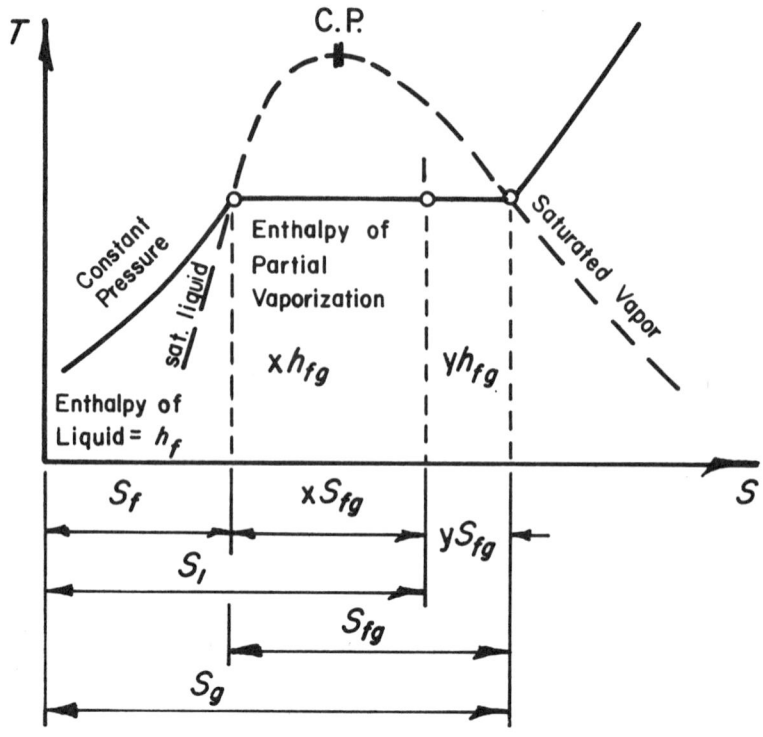

Fig. 7-1. Interpretation of Dryness Fraction.

(1) are given by the following expressions:

	Property	Value for Mixture of Quality x
(Eq. 7.5)	Specific Volume:	$v_1 = v_f + xv_{fg}$
(Eq. 7.6)	Specific Internal Energy:	$u_1 = u_f + xu_{fg}$
(Eq. 7.7)	Specific Enthalpy:	$h_1 = h_f + xh_{fg}$
(Eq. 7.8)	Specific Entropy:	$s_1 = s_f + xs_{fg}$

An alternate computation may, of course, be made in terms of percent moisture $= y = (1 - x)$.

Example 7-6

Steam flows through a turbine at a rate of 1.26 kg/s. The absolute pressure of steam entering the turbine is 2068 kPa at a temperature of

Ch. 7 THERMODYNAMICS AND HEAT TRANSFER 205

214°C. The entrance and exhaust velocity of the steam are essentially equal and there are no radiation losses. The exhaust steam is at an absolute pressure of 6.9 kPa and the "quality" is 81%. Calculate the power transmitted to the turbine. From steam tables:

$$h_1 = 2798 \text{ kJ/kg} \quad h_f = 162 \text{ kJ/kg} \quad h_{fg} = 2410 \text{ kJ/kg}$$

Applying Equations (7.7) and (7.4b):

$$h_2 = h_f + xh_{fg} = 162 + 0.81(2410) = 2115 \text{ kJ/kg}$$

$$q - P = \dot{m}[(h_2 - h_1) + \tfrac{1}{2}(\bar{V}_2{}^2 - \bar{V}_1{}^2) + g(z_2 - z_1)]$$

and assuming, as stated, that $\Delta \bar{V} = 0$; $\Delta z = 0$; $q = 0$, then:

$$-P/\dot{m} = (h_2 - h_1) = 2115 - 2798$$

$$P/\dot{m} = 683 \text{ kJ/kg}$$

$$P = 1.26(683) = 860.6 \text{ kJ/s}$$

$$P = 860.6 \text{ kW}$$

Example 7-7

A vertical insulated cylinder with a diameter of 300 mm contains 1.5 kg of water at 18°C, and a piston resting upon the water exerts a constant pressure of 0.8 MPa (absolute). Heat is transmitted to the water electrically at a rate of 550 W. How long will it take for the piston to rise a distance of 1100 mm? (Refer to Table 7-3.)

The initial specific volume of the water is, by interpolation:

$$v = 0.001001 \text{ m}^3/\text{kg},$$

so that initial volume

$$V_1 = 1.5(0.001001) = 0.0015 \text{ m}^3$$

The initial specific enthalpy is:

$$h = 75.5 \text{ kJ/kg}.$$

The addition of heat raises the temperature of the water and eventually generates steam. Since the pressure is constant the temperature will not exceed the saturation level, which for 0.8 MPa is 170.4°C. Under these conditions, the specific volumes of water and steam are, respectively, $v_f = 0.001115 \text{ m}^3/\text{kg}$ and $v_g = 0.2403 \text{ m}^3/\text{kg}$, from the same table. Let the dryness fraction of the steam and water mixture be denoted by

x. Then the specific volume of the mixture is obtained by noting that $v_{fg} = (v_g - v_f)$ and applying Equation (7.6). Thus: $v_{fg} = 0.2403 - 0.001115$; $v_{fg} = 0.2392$ m³/kg and $v = v_f + xv_{fg}$ from which:

$$v = 0.001115 + x(0.2392) \quad \text{m}^3/\text{kg}$$

But the increase in volume is to be:

$$\Delta V = \left\{(0.300)^2 \frac{\pi}{4}\right\} 1.100 = 0.07775 \text{ m}^3$$

Accordingly, the total final volume is:

$$V_2 = 0.07775 + 0.00112 = 0.07887 \text{ m}^3$$

Since $V = mv$, then:

$$V_2 = 0.07887 = 1.5\{0.001115 + x(0.2392)\}$$

which yields $x = 0.215$.

At 0.8 MPa and 170.4°C, the final specific enthalpy is obtained by noting that $h_{fg} = 2047$ kJ/kg and that $h = h_f + xh_{fg}$

$$h = 721 + 0.215(2047) \quad h = 1161 \text{ kJ/kg}$$

Hence, the increase in the enthalpy of the system is:

$$\Delta H = 1.5(1161 - 75.5) = 1628 \text{ kJ}$$

and since heat is being added at the rate of $q = 550$ W, the time taken for the desired expansion will be:

$$H = q \cdot t \quad t = 1628 \times 10^3/550 = 2960 \text{ s} = 2.96 \text{ ks}$$

which is $t = 49$ min and 20 seconds.

Note the absence of any special treatment of the electrical input; there is only one SI unit for energy!

The availability of a wide range of tables of thermodynamics properties of various fluids enables us to solve many other problems expressed in practical terms.

Example 7-8

Dry saturated steam is supplied to a turbine at 10 MPa (absolute), and 500°C, at the rate of 100 Mg/h. It enters the machine with negligible velocity, but leaves through a duct of cross-section 3.0 m² with a dryness fraction of 0.89 and a pressure of 7.0 kPa (absolute). What is the

Ch. 7 THERMODYNAMICS AND HEAT TRANSFER

velocity of the steam in the exhaust duct, and the power output of the turbine, assuming that heat losses are negligible?

At entry, $\bar{V}_1 = 0$; $\dot{m} = 100\,000/3600 = 27.8$ kg/s

and from reference tables, for 10 MPa and 500°C:

$$h_g = 3373 \text{ kJ/kg}; \quad v_g = 3.275 \times 10^{-2} \text{ m}^3/\text{kg}.$$

At exhaust, for saturated steam at 7.0 kPa, the values taken from references such as Table 7-5 are:

$$h_g = 2572 \text{ kJ/kg} \quad v_g = 20.53 \text{ m}^3/\text{kg}$$
$$h_f = 163 \text{ kJ/kg} \quad v_f = 0.001 \text{ m}^3/\text{kg}$$

noting that: $v_{fg} = v_g - v_f$, and utilizing Equation (7.5):

$$v_{fg} = 20.53 - 0.001$$
$$v_{fg} = 20.53 \text{ m}^3/\text{kg}$$
$$v = v_f + xv_{fg}$$
$$v_2 = 0.001 + 0.89(20.53)$$
$$v_2 = 18.27 \text{ m}^3/\text{kg}$$

In a similar manner, applying Equation (7.7), the enthalpy at exit is given by:

$$h_{fg} = 2572 - 163$$
$$h_{fg} = 2409 \text{ kJ/kg}$$
$$h = h_f + xh_{fg}$$
$$h_2 = 163 + 0.89(2409) = 2307 \text{ kJ/kg}$$

The exhaust velocity is given by:

$$\bar{V}_2 = \frac{\dot{m}v}{A}$$

$$\bar{V}_2 = \frac{100\,000}{3600} \times \frac{18.27}{3.0} \quad \left[\frac{\text{kg}}{\text{s}} \cdot \frac{\text{m}^3}{\text{kg}} \cdot \frac{1}{\text{m}^2} = \frac{\text{m}}{\text{s}}\right]$$

$$\bar{V}_2 = 169 \text{ m/s}$$

In summary, Equation (7.4b) for steady-flow energy, gives:

$$q - P = \dot{m}\{(h_2 - h_1) + \tfrac{1}{2}(\bar{V}_2^2 - \bar{V}_1^2)\}$$

$$q - P = 27.8\{(2307 - 3373)10^3 + \tfrac{1}{2}[(169)^2 - 0]\}$$

$$q - P = \frac{\text{kg}}{\text{s}}\left\{\frac{\text{J}}{\text{kg}} + \frac{\text{m}^2}{\text{s}^2}\right\}, \text{ but since } \frac{\text{m}^2}{\text{s}^2} = \frac{\text{J}}{\text{kg}}$$

therefore, $q - P$ is in terms of $\frac{\text{J}}{\text{s}} = \text{W}$. If heat losses are negligible, then $q = 0$, and $P = 29.6 \times 10^6$ W or, 29.6 MW.

Note that in calculations such as this one, where enthalpies determined from tables are added to kinetic energy terms calculated separately, care should be taken not to overlook the fact that the enthalpies are tabulated in kilojoules (kJ = 1000 J). Enthalpies are usually the largest components in the total energy.

Example 7-9

Steam flowing through a main at 500 kPa (absolute) is sampled and passed through a separating and throttling calorimeter. What is the dryness fraction of the steam in the main if the following results are obtained?

Mass of water separated, 15 g; mass of condensate collected, 0.5 kg; pressure after throttling, 100 kPa; temperature after throttling, 115°C.

From reference tables, at 500 kPa, $h_g = 2749$, $h_f = 640$, and $h_{fg} = 2109$ kJ/kg. Also the tables for 100 kPa and 115°C give $h_g = 2706$ kJ/kg.

Since the throttling process does not alter the enthalpy, assuming that kinetic energy does not change, an expression may be written in terms of an unknown dryness factor x' as follows:

$h = h_f + x'h_{fg}$ $2706 = 640 + x'(2109)$ from which $x' = 0.9796$

This, however, is after separation. To obtain the dryness fraction in the steam as sampled, the 15 g of water must be included with the appropriate quantity of steam.

After separation the mass of steam is: 0.5 kg, of which a fraction 0.9796 is steam, i.e., 0.4898 kg steam and 0.0102 kg water; hence, before separation there was 0.0102 + 0.015 = 0.0252 kg water. Accordingly the

Ch. 7   THERMODYNAMICS AND HEAT TRANSFER   209

original dryness fraction was:
$$x = 0.4898/(0.4898 + 0.0252)$$
$$x = 0.951$$

7.8 Entropy

Entropy is an intrinsic physical property of a closed thermodynamic system relating heat transferred to the thermodynamic temperature at which it is transferred. For any reversible process experienced by the system between States 1 and 2, the change in entropy is given by:

(Eq. 7.9) $\quad S_2 - S_1 = \int_1^2 \left(\dfrac{dQ}{T}\right)$, or $dS = \left(\dfrac{dQ}{T}\right)$ (for reversible process)

Entropy S, therefore, is effectively heat transferred divided by absolute temperature and is accordingly expressed in J/K, or in J/(kg·K) for, s, specific entropy. In tables, etc., entropy is usually given in terms of s = kJ/(kg·K).

The following additional relationships will be of value in working with the property of entropy:

(Eq. 7.10) $\quad Tds = du + pdv$

(Eq. 7.11) $\quad Tds = dh - vdp$

(Eq. 7.12) $\quad s_2 - s_1 = c_p \ln(T_2/T_1)$ $\quad$ (for constant pressure process)

(Eq. 7.13) $\quad s_2 - s_1 = c_v \ln(T_2/T_1)$ $\quad$ (for constant volume process)

in which c_p and c_v are the respective specific heat capacities in units of J/(kg·K), and are constant values for the process.

Further, it will be well to observe at this point that $s_2 = s_1$ for a reversible adiabatic process and similarly, that $s_2 > s_1$ for an irreversible adiabatic process.

The examples which follow will illustrate the use of these principles and their relationship to appropriate SI units.

Example 7-10

Dry, saturated steam flows steadily through an engine, entering at 700 kPa (absolute), and leaving at 40 kPa (absolute). What is the greatest possible output of work from the engine? Heat losses as well as entry and exit velocities are to be neglected for the purposes of this example.

From reference tables for steam:

	kJ/kg	kJ/(kg·K)
for: $p_s = 700$ kPa	$h_g = 2764$	$s_g = 6.709$
for: $p_s = 40$ kPa	$h_f = 318$	$s_f = 1.026$
	$h_g = 2631$	$s_g = 7.669$
	$h_{fg} = 2313$	$s_{fg} = 6.643$

For a given maximum temperature the maximum possible output will be achieved by a reversible engine from which the entropy would be the same at exit as at entry.

From this observation the first step would be to solve for the dryness fraction x which will give the entropy at exit equal to the entropy at entry.

Applying Equation (7.8):

$$1.026 + x(6.643) = 6.709, \quad \text{whence} \quad x = 0.855$$

Applying the computed value of $x = 0.855$, in Equation (7.7), the specific enthalpy in the exhaust steam is:

$$h = 318 + (0.855)2313 = 2295 \text{ kJ/kg}$$

Accordingly, the available output from such an engine is:

$$P = \dot{m}(h_2 - h_1)$$

$$\Delta h = 2764 - 2295 = 469 \text{ kJ/kg}$$

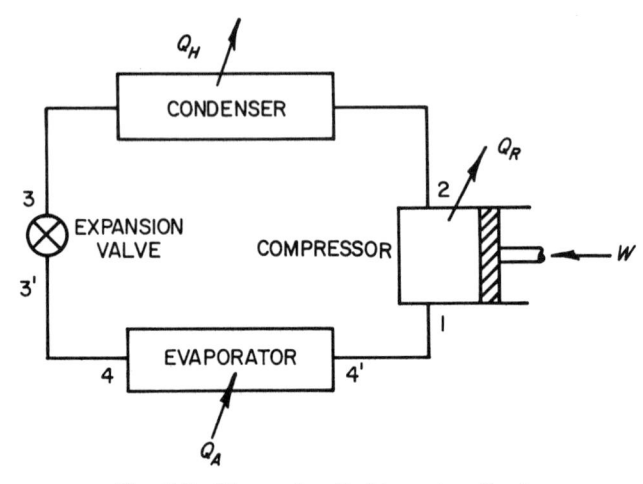

Fig. 7-2. Illustrative Refrigeration Cycle.

Ch. 7   THERMODYNAMICS AND HEAT TRANSFER   211

If multiplied by the mass flow rate in kg/s this would give the power output in watts.

Example 7-11

Measurements on a vapor compression refrigeration system (see Fig. 7-2) operating under steady conditions, for which the working medium is Refrigerant 12 (Freon 12), yield the following data:

Sta.	Location	Absolute Pressure kPa	Temperature °C
(1)	Entering compressor	140	−6
(2)	Leaving compressor	960	82
(3)	Entering expansion valve	930	37
(3')	Leaving expansion valve	180	(saturated mixture)
(4)	Entering evaporator	172	(saturated mixture)
(4')	Leaving evaporator	150	−12

$\dot{m}$ = mass flow rate of R-12 is: 130 kg/h; which converts to:
$\dot{m}$ = 0.0361 kg/s
P = electrical power input to compressor drive = 2000 watts
η = estimated efficiency of compressor drive = 97 percent

Calculate:

a) rate of heat transfer from the compressor to the surroundings (Q_R/m) in J/kg;
b) rate of heat transfer to the refrigerant in the evaporator region (Q_A/m) in J/kg; and
c) the coefficient of performance for the system (cop).

Solution: (see Fig. 7-3 for graphical representation)

a) Apply a modified statement of steady flow energy equation (Eqs. [7.2] and [7.3]) to the compressor:

$$(Q_R/\dot{m}) = (h_2 - h_1) + (P/\dot{m}) \qquad \frac{J}{kg} + \frac{J}{s} \cdot \frac{s}{kg} = \frac{J}{kg}$$

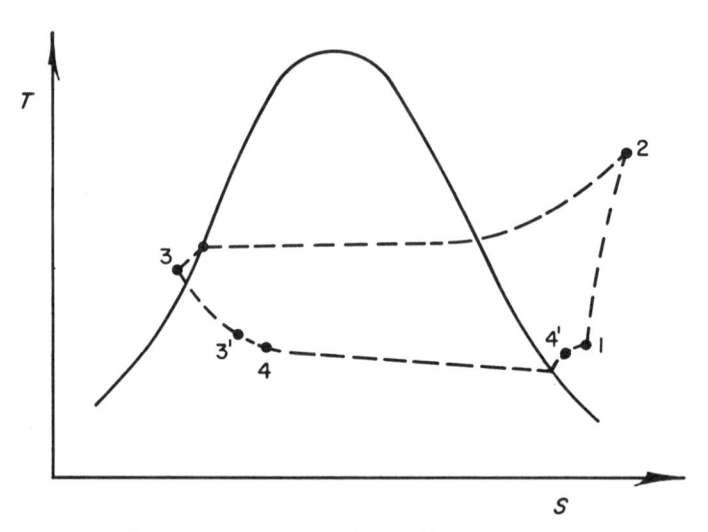

Fig. 7-3. *T-S* diagram for Refrigeration Cycle.

Using enthalpy values obtained from a table or chart for R-12,

$$(Q_R/m) = (234\,760 - 187\,110) - \frac{(0.97 \times 2000)}{0.0361}$$

$$= 47\,650 - 53\,740$$

$Q_R/m = -6\,090$ J/kg (the minus sign denoting heat transfer *from* the refrigerant *to* the surroundings.)

$q = \dot{m}(Q_R/m) = 0.0361(-6\,090) = 220$ J/s $= 220$W

b) The steady flow energy equation is now applied to the evaporator region. Neglecting heat transfer between the entrance to the expansion valve and the entrance to the evaporator ($h_4 = h_3$ for adiabatic throttling between these points):

$$(Q_A/m) = (h_{4'} - h_4) = (h_{4'} - h_3)$$

Again, obtaining the enthalpy values from a table or chart for R-12:

$(Q_A/m) = (183\,400 - 71\,360) = 112\,040$ J/kg

$q = \dot{m}(Q_A/m) = 0.0361(112\,040) = 4045$ J/s (where the plus sign denotes heat transfer *from* the surroundings *to* the refrigerant)

Ch. 7   THERMODYNAMICS AND HEAT TRANSFER   213

c) The coefficient of performance (cop) is determined by:

$$\text{cop} = \frac{\text{refrigeration effect}}{\text{electrical work input}}$$

$$\text{cop} = \frac{4045}{2000} = 2.02$$

Such calculations as the two just given are considerably facilitated by the presentation of the data in graphical form instead of in tables. These charts are, of course, available in SI units. Thus the temperature-entropy chart is normally expressed in K and kJ/(kg·K), and it will be noted that areas on this type of chart are in terms of $(K)\frac{kJ}{kg \cdot K}$, i.e., $\frac{kJ}{kg}$. The enthalpy-entropy chart is usually called the Mollier diagram, where the enthalpy is expressed in kJ/kg and the entropy in kJ/kg·K. Plotted on the enthalpy-entropy coordinates are constant pressure and constant temperature lines which coincide in the wet region; in the superheat region the constant temperature lines diverge from the constant pressure lines. In addition, lines of constant quality and constant superheat are shown.

7.9 Properties of Gases

The first and second laws of thermodynamics lead to a great variety of relations between the properties of gases, and it may be instructive to consider the SI units by which these are expressed.

As an illustrative example, the specific heat capacity of a gas depends upon the conditions imposed. It is customary to distinguish two particular cases, namely constant pressure and constant volume. The units of both are J/(kg·K). The definitions may usefully be examined from the "units" point of view:

(Eq. 7.14) $c_v = \left.\frac{du}{dT}\right|_v$ $\frac{J}{kg \cdot K} = \frac{J/kg}{K}$

(Eq. 7.15) $c_p = \left.\frac{dh}{dT}\right|_p$ $\frac{J}{kg \cdot K} = \frac{J/kg}{K}$

The physical properties of gases which are at low pressure compared to the critical pressure, behave in a much more regular way and conform, with sufficient accuracy for many purposes, to the simple relationship shown in the following equation:

(Eq. 7.16) $\qquad pv = RT, \quad \text{or,} \quad pV = mRT$

Analyzing this expression to obtain the SI units for R yields:

$$R = pv/T \qquad \frac{N}{m^2} \cdot \frac{m^3}{kg} \cdot \frac{1}{K} = N \cdot m/(kg \cdot K) = J/(kg \cdot K)$$

Each gas has its own value for the gas constant: for air it is 0.287 kJ/(kg·K).

The universal gas constant introduces the concept of the mole, a unit for measuring the "amount of substance"; it is that amount of substance whose mass is numerically equal to its molecular weight.

The official definition as decided by the 14th CGPM in 1971 is:

> ... The mole is the amount of substance of a system which contains as many elementary entities as there are atoms in 0.012 kilogram of carbon 12; its symbol is 'mol'.

Thus, in SI terms the mole (mol) is the gram-mole. For SI:

$$\text{gram molecular weight} = \frac{\text{grams of a substance}}{\text{molecular weight of substance}}$$

If the Equation $pv = RT$ is written for a mole of a gas instead of a kilogram, then v represents the volume of one mole, and the unit of the corresponding gas constant, R, is J/(mol·K); it is found that this form of gas constant has the same value for all gases, namely, 8.314 J/(mol·K).

Another useful figure is that the volume of one mole of an ideal gas at 0°C and 101.3 kPa (i.e., atmospheric pressure) is 0.02241 m³, regardless of the gas.

Specific heat capacities may also be defined on a molar basis, when the units become J/mol·K. It might be shown that for a gas obeying $pv = RT$:

(Eq. 7.17) $\qquad c_p - c_v = R \qquad \text{(ideal gas)}$

and it will be seen from the above that the units are consistent, whether expressed per unit mass (in which case the result R, is the specific gas constant), or whether expressed per mole (in which case the result is R_u the universal gas constant). For the ideal gas, $pv = RT$ and c_p and c_v are functions of temperature only.

For gases which conform sufficiently closely to these requirements a range of formulae have been derived for determining the enthalpy,

Ch. 7   THERMODYNAMICS AND HEAT TRANSFER   215

entropy, etc., obviating the need for tables or diagrams. These introduce no new concepts from the units point of view. The following repeats Example 7-9, but assumes that the working fluid is an ideal gas.

Example 7-12

An ideal gas at 700 kPa and 165°C flows steadily through an engine and emerges at 40 kPa. What is the greatest possible output of work? Heat losses, and entry and exit velocities are negligible. Assume $\gamma = 1.4$.

The maximum output is achieved when the process is reversible, i.e., isentropic. For such a process, with an ideal gas, it may be shown that:

(Eq. 7.18) $$pv^{\gamma} = \text{constant}$$

in which

(Eq. 7.19) $$\gamma = c_p/c_v,$$

the ratio of the two specific heats, and where the work done is given by:

(Eq. 7.20) $$W/m = \frac{\gamma(p_1 v_1 - p_2 v_2)}{(\gamma - 1)} \qquad \frac{\text{Pa·m}^3}{\text{kg}} = \frac{\text{N}}{\text{m}^2} \cdot \frac{\text{m}^3}{\text{kg}} = \frac{\text{N·m}}{\text{kg}} = \frac{\text{J}}{\text{kg}}$$

where p_1, v_1 and p_2, v_2 are, respectively, the entry and exit pressures and specific volumes. It will be assumed that for this gas $R = 0.3$ kJ/(kg·K). Then initially: $p_1 v_1 = RT_1$.

$$(700 \times 10^3)v_1 = 0.3 \times 10^3(273 + 165) \qquad \frac{\text{N}}{\text{m}^2} \cdot \frac{\text{m}^3}{\text{kg}} = \frac{\text{J}}{\text{kg·K}} \cdot \text{K} = \frac{\text{J}}{\text{kg}}$$

$$v_1 = 0.188 \text{ m}^3/\text{kg}$$

To obtain v_2 we note that:

$$p_1 v_1^{\gamma} = p_2 v_2^{\gamma}$$

$$(700 \times 10^3)(0.188)^{1.4} = (40 \times 10^3)v_2^{1.4}$$

whence

$$v_2 = 0.188(700/40)^{1/1.4}$$

$$v_2 = 1.45 \text{ m}^3/\text{kg}$$

By substitution in Equation (7.20) above, the work done is:

$$W/m = 1.4 \left[\frac{(700 \times 10^3)0.188 - (40 \times 10^3)1.45}{(1.4 - 1.0)} \right] = 257.5 \text{ kJ/kg}$$

A considerable variety of processes may be represented by formulae of the general type pv^γ = constant, where γ has an appropriate value; they are known as "polytropic." The use of the index γ seems to lead to a very peculiar unit, but this is really only an arithmetical manipulation, and the apparent difficulty disappears if one writes $(v_1/v_2)^\gamma$, since (v_1/v_2) results in neutral dimensions.

Example 7-13

A fuel of gravimetric analysis 87% (C), 12% (H_2), 1% (O_2), is burned with 70% air in excess of the stoichiometric requirement. Find the mass of air supplied per kilogram of fuel.

The analysis shows that a kilogram of fuel consists of: 0.87 kg (C), 0.12 (H_2), 0.01 (O_2); the molecular masses are, respectively: 12, 2, and 32; hence the composition in molar terms is:

0.87/12 kmol (C); 0.12/2 kmol (H_2), 0.01/32 kmol (O_2).

Note the use of the kilomole, kmol, i.e., a thousand moles; this is because the masses are expressed in kilograms, whereas the mol is a gram-mole.

Let the stoichiometric quantity of air required be x kmol of air per kg of fuel. From the known composition of air the stoichiometric equation (in terms of the kmol) will then be:

$$\frac{0.87}{12}(C) + \frac{0.12}{2}(H_2) + \frac{0.01}{32}(O_2) + 0.21x\,(O_2)$$
$$+ 0.79x\,(N_2) = a\,(CO_2) + b\,(H_2O) + c\,(N_2)$$

where a, b, and c have yet to be determined, since with complete combustion no free oxygen, carbon or hydrogen will be left. This equation must be correct for each of the molecules involved. Thus it follows that:

(C) = 0.87/12 = a; (H_2) = 0.12/2 = b; (O_2) = (0.01/32) + 0.21x = a + (b/2)

whence

$$\frac{0.01}{32} + 0.21x = \frac{0.87}{12} + \frac{0.12}{4}$$

so that

$$x = 0.487$$

The stoichiometric amount of air per kilogram of fuel is therefore

0.487 kmol, which has a mass of 0.487(0.21 × 32 + 0.79 × 28) = 14.0 kg. A 70% excess will therefore require 14.0(1.70) = 23.8 kg of air.

7.10 Modes of Heat Transfer

While heat transfer conditions in actual practice usually represent a combination of several different phenomena, there is considerable merit, from a notation and unit system point of view, in reviewing these occurrences initially as three independent basic modes. With a clear concept of each mode (conduction, convection, and radiation) the appropriate combinations can readily be made, where required, in later applications.

7.11 Conduction

Thermal conduction occurs through an element when a temperature differential exists between two faces or interfaces of the element. The amount of energy transferred in the form of heat is dependent upon the temperature gradient, on the length and cross-sectional area of the path, and on a factor termed the "thermal conductivity," a physical characteristic of the material through which the transfer is occurring. The expression for the relationship of these factors is generally known as Fourier's Law:

(Eq. 7.21) $\quad q = k \cdot A \cdot \Delta T / \Delta x \quad \dfrac{W}{m \cdot K} \cdot m^2 \cdot \dfrac{K}{m} = W = \dfrac{J}{s}$

in which:

q = heat transfer rate .... J/s = W
A = cross section of transfer path .... m²
ΔT = temperature differential .... K
Δx = length of path, usually thickness of element through which heat is being transferred .... m
k = thermal conductivity .... W/(m·K)

Some representative values of k are given in Table 7-6. Note that conductivity generally increases with the level of absolute temperature. For actual conditions a more complete reference should be consulted. Conversions from customary units can be made by observing that:

$\quad\quad$ 1.0000 Btu/(hr·ft·°F) = 1.7307 W/(m·K)

218   THERMODYNAMICS AND HEAT TRANSFER   Ch. 7

Table 7-6. Thermal Conductivity Coefficients k (@300K) W/(m·K)

Material	Thermal Cond., k	Material	Thermal Cond., k	Material	Thermal Cond., k
Aluminum	202	Glass	0.8–1.1	Water	0.614
Copper	386	Brick	0.4–0.7	Steam	0.184
Steel	55	Concrete	0.9–1.4	Air	0.0262

Example 7-14

A steel plate 25 mm in thickness is subjected to a temperature differential of 40 K between faces. What is the heat transferred from face to face of the plate if the magnitude of the thermal conductivity is: k = 55 W/(m·K)?

Applying Fourier's Law (Equation 7.21) for heat transfer by conduction:

$$q = kA \cdot \Delta T / \Delta x$$

the heat transferred per unit of area is:

$$q = \frac{55(1.0)40}{0.025} = 88\,000 \text{ W/m}^2 = 88.0 \text{ kW/m}^2$$

which may also be expressed as:

$$88.0 \text{ kJ/(s·m}^2\text{)}, \quad \text{or,} \quad 317 \text{ MJ/(h·m}^2\text{)}$$

7.12 Convection

When a temperature differential exists between a solid and a fluid with which it interfaces, there is a phenomenon termed "convection" which develops in a boundary layer (film) at the surface of contact. The heat transferred per unit of time is a function of the film conductance, the

Table 7-7. Surface Heat Transfer Coefficient h W/(m²·K)

AIR	Approx. Range	WATER	Approx. Range
By free convection	5–25	By forced convection	10–10 000
By forced convection	10–500	With boiling	2000–10 000
		With condensation	10 000–100 000

area of contact, and the temperature differential. The result, in terms of Newton's Law of Cooling, may be expressed as:

(Eq. 7.22) $\quad q = hA(T_1 - T_2) \qquad \dfrac{W}{m^2 \cdot K} \cdot m^2 \cdot K = W = \dfrac{J}{s}$

in which:

q = heat transfer rate .... J/s = W
A = surface area of contact .... m^2
T_1 = temperature of solid at face .... K
T_2 = temperature of fluid beyond film .... K
h = surface heat transfer coefficient .... W/($m^2 \cdot$K)
(also called "film coefficient")

As would be expected, a temperature differential is coupled with a fluid flow at the interface. Thus, the resultant heat transfer rate is dependent on the laws of fluid dynamics as well as on the thermal characteristics of the fluid. Hence, the film coefficient, h, is a function of fluid properties such as viscosity and density, as well as thermal properties such as specific heat and thermal conductivity of the fluid.

In addition, a distinction must be made between the condition of natural or "free" convection in which the main body of the fluid is relatively at rest, in contrast with the condition of "forced convection" where an external means is engaged to move the fluid stream across the interface at a rate greater than that which would occur naturally.

Under certain conditions, as in a heat exchanger tube, a value for the convection heat-transfer coefficient may be calculated from an empirical equation such as:

$$\dfrac{hD}{k} = n \left(\dfrac{\bar{V}D\rho}{\mu}\right)^a \left(\dfrac{c\mu}{k}\right)^b$$

Note the involvement of Reynolds number (Eq. 6.22) and the Prandtl number ($c\mu/k$). In such an expression when the h is in units coherent with D/k, as in SI, then the factor of proportionality, n, is dependent only on experimental data and is free of dimensional characteristics.

Some idea of the approximate range of values of convection heat transfer coefficients is given in Table 7-7. Values of convection coefficients in customary units may be converted by noting that:

$$1.0000 \text{ Btu/(hr} \cdot \text{ft}^2 \cdot \text{°F)} = 5.6783 \text{ W/(m}^2 \cdot \text{K)}$$

220 THERMODYNAMICS AND HEAT TRANSFER   Ch. 7

Example 7-15

Water at 80 C° is pumped across a metal surface area of 3.6 square meters which is maintained at 150 C°. If the surface heat transfer coefficient is 500 W/(m²·K), what will be the heat transferred by convection?

Applying Eq. (7.22):

$$q = h \cdot A(T_1 - T_2)$$

$$q = 500 \times 3.6(150 - 80) = 126\,000 \text{ W} = 126 \text{ kW}$$

$$\text{or,} \quad 126 \text{ kJ/s} = 453.6 \text{ MJ/h}$$

7.13 Radiation

The mechanism of heat transfer by radiation is electromagnetic in character and occurs even in a vacuum. From a thermal point of view, the potential of a body to radiate energy is a function of the fourth power of its absolute temperature. Usually the interest is in the heat exchanged between two bodies. A general expression, based on the Stefan-Boltzmann Law may be written as:

(Eq. 7.23) $\quad q = F_G F_\epsilon \sigma A(T_1^4 - T_2^4) \quad \dfrac{\text{W}}{\text{m}^2 \cdot \text{K}^4} \cdot \text{m}^2 \cdot \text{K}^4 = \text{W} = \text{J/s}$

in which:

q = heat transfer rate $\quad$ J/s = W
A = area of radiating surface $\quad$ m²
T = absolute temperature $\quad$ K
F_G = geometric (view) function depending on shape and extensiveness of surfaces exchanging radiation
F_ϵ = emissivity function
σ = Stefan-Boltzmann radiation factor = 5.670×10^{-8} W/(m²·K⁴)

Emissivity is the ratio of the emissive power of a given body to that of a black body at the same temperature. As a ratio ϵ has no units. Some representative values of ϵ are given in Table 7-8.

Example 7-16

A large polished steel plate with a surface temperature maintained at 80°C is separated by a narrow air space of uniform width from a pool of

Ch. 7   THERMODYNAMICS AND HEAT TRANSFER   221

Table 7-8. Normal Emissivities ϵ (at 300 K)

Material	Normal Emissivities, ϵ	Material	Normal Emissivities, ϵ
Aluminum			
(polished)	0.04	White paint	0.90
(oxidized)	0.2-0.31	Plaster	0.92
Steel			
(polished)	0.07	Brick	0.93
(oxidized)	0.79	Water	0.96

water which has a constant temperature of 20°C. What is the likely heat transfer by radiation between the two surfaces per unit of exposed area?

In order to proceed with this solution it is first necessary to determine the absolute temperatures:

$$T_1 = 80 + 273 = 353 \text{ K}; \quad T_2 = 20 + 273 = 293 \text{ K}$$

For this arrangement of extensive parallel surfaces which "see" only each other, $F_G = 1.0$, and the emissivity function F_ϵ is given by:

$$F_\epsilon = \frac{1}{\dfrac{1}{\epsilon_1} + \dfrac{1}{\epsilon_2} - 1} = \frac{1}{\dfrac{1}{0.07} + \dfrac{1}{0.96} - 1} = 0.0698$$

Applying the Stefan-Boltzmann Equation, Eq. (7.23)

$$q/A = 1.0(0.0698)(5.670 \times 10^{-8})[(353)^4 - (293)^4]$$

$$q/A = 32.3 \text{ W/m}^2$$

7.14 Overall Heat Transfer

When heat transfer occurs through a composite element, such as the wall of a building, a sequence of conduction and convection coefficients may be involved. As in other "series" type problems, the approach to determining the combined or "overall" coefficient U is based on a summing of the resistances, which is the sum of the reciprocals of the conductances in the path of heat transfer.

For this purpose, the following definitions may be applied to identify

222   THERMODYNAMICS AND HEAT TRANSFER   Ch. 7

the coefficients applicable to conductance:

K = thermal conductance;

$$K = \frac{kA}{L} \qquad \frac{W}{m \cdot K} \cdot \frac{m^2}{m} = W/K$$

R = thermal resistance;

$$R = \frac{L}{kA} \qquad \frac{m \cdot K}{W} \cdot \frac{m}{m^2} = K/W$$

Frequently the above factors may be stated in terms of unit areas. A complete check on the units of any data taken from reference tables is therefore advisable. Convection and radiation coefficients may be treated in a similar manner.

The relationships for overall heat transfer may be stated as:

(Eq. 7.24) $$q = U \cdot A \cdot \Delta T$$

in which:

q = heat transfer rate .... W = J/s
A = cross-sectional area of heat transfer path ...... m²
ΔT = overall temperature differential .... K
U = overall heat transfer coefficient .... W/(m²·K)

To determine U it is frequently necessary to use the relationship:

(Eq. 7.25) $$\frac{1}{U} = R_1 + R_2 + R_3, \text{ etc.} \qquad \frac{1}{U} = R_T \qquad \frac{W}{m^2 \cdot K}$$

This may alternately be stated as:

(Eq. 7.26) $$R_T = \frac{1}{h_1} + \frac{L_2}{k_2} + \frac{L_3}{k_3} + \frac{L_4}{k_4} + \frac{1}{h_2}, \text{ etc.}$$

$$\frac{m^2 \cdot K}{W} + (m) \frac{m \cdot K}{W}, \text{ etc.} = \frac{m^2 \cdot K}{W}$$

Example 7-17

An exterior building wall consists of 100 mm of brick, 150 mm of dense concrete, and 20 mm of gypsum plaster for which the thermal conductivities, respectively, are: k = 0.50, 1.50, and 1.20 W/(m·K). The surface heat transfer (film) coefficients are: (interior) h_i = 8.10, and

Table 7-9. Fuel Values and Equivalents

Heat Source	Unit	Net Heating Value, (MJ)	t, Coal	Heating Oil t	Heating Oil m³	Heating Oil bbl	Gas, 1000 m³	Electricity, 1000 kWh
Coal	t	29 400	1	0.70	0.78	4.9	0.84	8.14
Heating Oil	t	42 000	1.43	1	1.11	7.0	1.20	11.63
" "	m³	37 800	1.29	0.9	1	6.3	1.08	10.47
" "	bbl	6 000	0.204	0.143	0.159	1	0.172	1.66
Natural Gas	1000 m³	34 900	1.19	0.83	0.92	5.82	1	9.65
Electricity	1000 kWh	3 600	0.123	0.086	0.0955	0.60	0.104	1

224 THERMODYNAMICS AND HEAT TRANSFER Ch. 7

(exterior) $h_e = 19.0$ W/(m²·K). What is the heat loss through a 3.0 m by 7.0 m panel of this wall when there is a temperature differential of 40 K?

Applying Equations (7.26), (7.25), and (7.24), the unit thermal resistance is:

$$R_T = \frac{1}{8.10} + \frac{0.100}{0.50} + \frac{0.150}{1.50} + \frac{0.020}{1.20} + \frac{1}{19.0}$$

$R_T = 0.4928$ m²·K/W

$U = 1/R_T$ $U = 2.029$ W/(m²·K)

$q = U \cdot A \cdot \Delta T$ $q = 2.029(3.0 \times 7.0)40$

$q = 1704$ W $q = 1.704$ kW

7.15 Heating Value of Fuels

When the quantity of heat from all sources is expressed in a common, coherent unit, the joule (J), many useful direct comparisons can readily be made as is illustrated in the table of approximate values, Table 7-9.

8

Electricity, Magnetism, and Light

Electrical science and electrical engineering have been associated with SI and the roots of SI for many years. For this reason it is highly instructive to take a more intense look at the definition of units and the beneficial evolution of the unit system in the electrical world.

At the core of this association of electrical science with SI is the definition of the ampere (A) which is based on the long known fact that a current flowing through a conductor produces a magnetic effect in the space around the conductor. Thus, if there are *two* parallel conductors, each with a current, a *force* will be generated between them. In rationalized form Ampere's Law states that the resultant force F is given by the equation:

$$F = \mu\, I_1 \cdot I_2 \cdot L / 2\pi r$$

where μ is the permeability of space, I_1 and I_2 the respective currents, L the parallel length of the conductors, and r the distance between the conductors. Many actual measurements have been made to evaluate this force and a precision of a few parts per million has now been achieved.

From many years of imaginative and enlightened work by the International Electrotechnical Commission (IEC), there has emerged the SI definition of the ampere as portrayed in Fig. 8-1.

The full official wording of this definition of the ampere is given in a later section with the description of other mechanical and electrical units.

An interesting sidelight is that a "force" definition of the ampere was originally proposed by Professor Giorgi in the year 1901, was first adopted by IEC around 1940, and was accepted by CGPM in formulating SI in 1960.

A further step taken in formulating SI was to declare the value of the magnetic constant, usually called the permeability factor, as:

$$\mu = 4\pi \times 10^{-7} \text{ H/m},$$

226   ELECTRICITY, MAGNETISM, AND LIGHT   Ch. 8

$I_1 = 1.0$ ampere $= 1.0$ A

$r = 1.0$ m

$F =$ force $= 2.0 \times 10^{-7}$ N

parallel, infinite conductors; negligible size; in vacuum

$\mu = 4\pi \times 10^{-7}$ H/m

$L = 1.0$ m

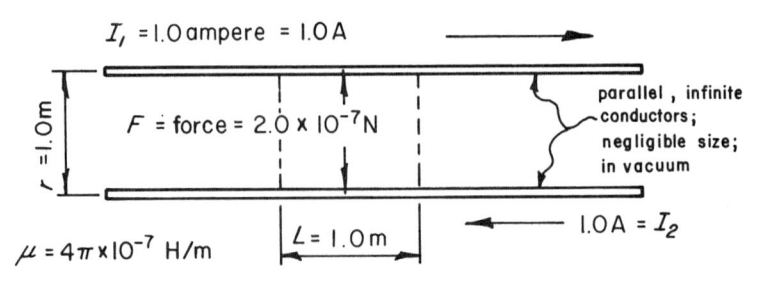

1.0 A $= I_2$

Fig. 8-1. The Ampere.

thus making the electromagnetic part of SI equivalent to the MKSA, or rationalized Giorgi system of IEC. SI thus assumes that all electromagnetic quantities will be in standard rationalized units and that all electrical units will thereby be coherently related to mechanical units.

Earlier procedure for evaluating electric current was through electrochemical phenomena. As a result the "international ampere" was defined as that "unvarying current which in one second deposits 0.001 118 gram of silver from an aqueous solution of silver nitrate (AgNO$_3$)." Note that there was involvement of mass and time in this previous definition but no mention of force, length, or distance.

Later, the "absolute ampere" (just slightly larger than the original international ampere) became the most universally used unit of electric current.

In more recent times, in the "centimeter, gram, second, electromagnetic-unit (cgs-emu) system, still another unit, the "abampere" has been the unit of electric current. In magnitude the abampere corresponds to ten times the absolute ampere. It is the definition of the abampere which represented the first major official step toward the present force concept. According to that definition the parallel conductors were to be placed one centimeter apart and the resultant force between the conductors was to be two dynes per centimeter of length. Comparison with the present SI definition may be made by recognizing that the newton now used as the unit of force is 10^5 dyne, that the distances referred to are now in terms of the meter rather than a centimeter, and that the value of the permeability factor has been adjusted to provide a rationalized, coherent formula.

Another aspect of the evolutionary process was the concept of the "statampere" which is a part of the "centimeter, gram, second, electrostatic-unit" (cgs-esu) system. In magnitude the "statampere" is a very small unit since it takes 2.998×10^9 of such units to equal one absolute ampere. Note the numerical association in this instance with the speed of light expressed in meters per second.

As has been stated, the SI definition of the ampere in terms of force, provides a favorable direct linkage between electrical and mechanical energy and accordingly establishes a common basis for all "power ratings." From a "units" point of view entirely, it may be stated that:

$$N \cdot m = J \qquad J/s = W \qquad J = W \cdot s \qquad W = A \cdot V \qquad J = A \cdot V \cdot s$$

Thus, in a world on the brink of many new developments in the utilization of energy, SI provides a vital new key to total systems concepts.

The importance of this relationship in the minds of the authors of SI is demonstrated by the fact that the ampere (A) was declared a Base Unit, notwithstanding the fact that its definition is dependent on a Derived Unit, the newton. By this declaration it is made clear that the ampere is the central connecting point between the electrical unit system and those of all other sciences.[1]

In view of the essential nature of this connection it is advisable to review the exact words of CIPM (1946) and CGPM (1948) when these historic actions were taken. The key definitions are:

A) Definitions of the mechanical units which enter the definitions of electric units:

Unit of force.—The unit of force (in the MKS [meter, kilogram, second] system) is that force which gives to a mass of one kilogram an acceleration of one meter per second squared. (This is now called the newton.)

Joule (unit of energy or work).—The joule is the work done when the point of application of one MKS unit of force (newton) moves a distance of one meter in the direction of the force.

Watt (unit of power).—The watt is that power which in one second gives rise to energy of one joule.

---

[1] For a more complete analysis of these key evolutionary developments in unit system concepts see: "Standard Handbook of Electrical Engineering," 10th ed., edited by Fink and Carroll, Chapter 1, by Bruce B. Barrow; also see historical references concerning the many imaginative contributions by Lord Kelvin.

B) Definitions of electric units. The CIPM accepts the following propositions which define the theoretical value of the electric units:

Ampere (unit of electric current).—The ampere is that constant current which, if maintained in two straight parallel conductors of infinite length, of negligible circular cross section, and placed one meter apart in a vacuum, would produce between these conductors a force equal to 2×10^{-7} MKS unit of force (newton) per meter of length.

Volt (unit of potential difference and of electromotive force).—The volt is the difference of electric potential between two points of a conducting wire carrying a constant current of one ampere, when the power dissipated between these points is equal to one watt.

Ohm (unit of electric resistance).—The ohm is the electric resistance between two points of a conductor when a constant potential difference of one volt, applied to these points, produces in the conductor a current of one ampere, the conductor not being the seat of any electromotive force.

Coulomb (unit of quantity of electricity).—The coulomb is the quantity of electricity carried in one second by a current of one ampere.

Farad (unit of electric capacitance).—The farad is the capacitance of a capacitor between the plates of which there appears a potential difference of one volt when it is charged by a quantity of electricity of one coulomb.

Henry (unit of electric inductance).—The henry is the inductance of a closed circuit in which an electromotive force of one volt is produced when the electric current in the circuit varies uniformly at the rate of one ampere per second.

Weber (unit of magnetic flux).—The weber is that magnetic flux, which linking a circuit of one turn, would produce in it an electromotive force of one volt if it were reduced to zero at a uniform rate in one second.

From these descriptions it will be noted that the unit definitions of the volt, ohm, coulomb, farad, and henry remain practically unchanged. The weber replaces the maxwell as the unit of magnetic flux.

From other references it will be noted that CIPM considers it advisable *not* to use with SI any of the units of the so-called "electro-

Ch. 8   ELECTRICITY, MAGNETISM, AND LIGHT   229

magnetic" 3-dimensional (MKS) system which, strictly speaking, cannot be compared to the corresponding unit of the 4-dimensional (MKSA) system now known as SI. This includes the oersted and the gauss; the latter is now succeeded by the tesla.

The most recent of new units with special names, in the electrical category, is the siemens which succeeds the mho. For more comprehensive details of these interrelationships and appropriate conversion factors, see Table 8-1.

A graphical representation of the coherent relationships in SI between selected Base Units and some of the key Derived Units in electricity and magnetism is given later in Chapter 9, Fig. 9-5.

Light

The candela (cd), sometimes called "new candle" by action of CGPM in 1948, replaced the earlier units of luminous intensity based on flame or incandescent filament standards. The candela (cd) was later incorporated in SI as a Base Unit.

This new international standard of light is based on the luminous intensity of a theoretically perfect radiation source (blackbody) which glows incandescently at a particular high temperature. Specifically the definition states:

> The candela is the luminous intensity, in the perpendicular direction, of a surface of 1/600 000 square meter of blackbody at the temperature of freezing platinum under a pressure of 101 325 newtons per square meter.

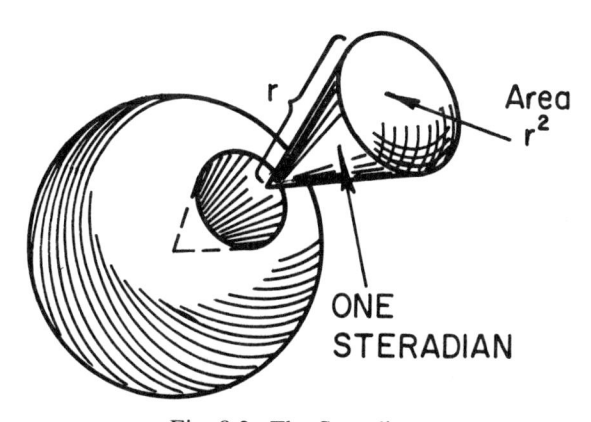

Fig. 8-2. The Steradian.

Table 8-1. Electricity and Magnetism[21]

To convert from	to	Multiply by
abampere	ampere (A)	1.000 000 * E+01
abcoulomb	coulomb (C)	1.000 000 * E+01
abfarad	farad (F)	1.000 000 * E+09
abhenry	henry (H)	1.000 000 * E−09
abmho	siemens (S)	1.000 000 * E+09
abohm	ohm (Ω)	1.000 000 * E−09
abvolt	volt (V)	1.000 000 * E−08
ampere hour	coulomb (C)	3.600 000 * E+03
EMU of capacitance	farad (F)	1.000 000 * E+09
EMU of current	ampere (A)	1.000 000 * E+01
EMU of electric potential	volt (V)	1.000 000 * E−08
EMU of inductance	henry (H)	1.000 000 * E−09
EMU of resistance	ohm (Ω)	1.000 000 * E−09
ESU of capacitance	farad (F)	1.112 650 E−12
ESU of current	ampere (A)	3.335 6 E−10
ESU of electric potential	volt (V)	2.997 9 E+02
ESU of inductance	henry (H)	8.987 554 E+11
ESU of resistance	ohm (Ω)	8.987 554 E+11
faraday (based on carbon-12)	coulomb (C)	9.648 70 E+04
faraday (chemical)	coulomb (C)	9.649 57 E+04

Ch. 8 ELECTRICITY, MAGNETISM, AND LIGHT

faraday (physical)	coulomb (C)	9.652 19 E+04
gamma	tesla (T)	1.000 000* E−09
gauss	tesla (T)	1.000 000* E−04
gilbert	ampere (A)	7.957 747 E−01
maxwell	weber (Wb)	1.000 000* E−08
mho	siemens (S)	1.000 000* E+00
oersted	ampere per meter (A/m)	7.957 747 E+01
ohm centimeter	ohm meter ($\Omega \cdot$m)	1.000 000* E−02
ohm circular-mil per foot	ohm millimeter^2 per meter ($\Omega \cdot$mm^2/m)	1.662 426 E−03
statampere	ampere (A)	3.335 640 E−10
statcoulomb	coulomb (C)	3.335 640 E−10
statfarad	farad (F)	1.112 650 E−12
stathenry	henry (H)	8.987 554 E+11
statmho	siemens (S)	1.112 650 E−12
statohm	ohm (Ω)	8.987 554 E+11
statvolt	volt (V)	2.997 925 E+02
unit pole	weber (Wb)	1.256 637 E−07

[21] ESU means electrostatic cgs unit. EMU means electromagnetic cgs unit. "Metric Standards for Engineering," *British Standard Handbook* No. 18, 1972.

* An asterisk after the sixth decimal place indicates that the conversion factor is exact and that all subsequent digits are zero. All other conversion factors have been rounded to the figures given in accordance with ANS Z210.1-1976 (ASTM E 380-76; IEEE Std. 268-1976). Sec. 4.4: "Rounding Values." Where less than six decimal places are shown, more precision is not warranted.

Table 8-2. Light

To convert from	to	Multiply by
footcandle	lux (lx)	1.076 391 E+01
footlambert	candela per meter^2 (cd/m^2) ..	3.426 259 E+00
lambert	candela per meter^2 (cd/m^2) ..	3.183 099 E+03

The SI unit of luminous flux is the lumen (lm). A source having an intensity of one candela in all directions radiates a luminous flux of 4π lumens. More specifically, the exact definition says that a "lumen" is:

> The luminous flux emitted in a solid angle of one steradian by a uniform point source having an intensity of one candela.

Hence the need for recognition officially in SI of the steradian (see Fig. 8-2), which is classified as a "supplementary unit" and defined as:

> ... the solid angle with its vertex at the center of a sphere that is subtended by an area of the spherical surface equal to that of a square with sides equal in length to the radius.

As a bench mark of "equivalence," a 100-watt light bulb can be said to emit about 1700 lumens.

The SI unit of "illuminance" is the lux (lx) which is lumens per square meter.

Conversion factors for some units of illuminance and luminance are given in Table 8-2.

9

Conversion to Preferred SI Usage

9.1 To Comprehend the System

At the outset of this volume it was stated that initial emphasis should be wholly on comprehending SI as a new, coherent, simplified, absolute *system* of measuring units. It also has been very fittingly said that a prime purpose of "Going SI" is the "de-Babelization" of the world of measurement. As a new worldwide language, SI can achieve that goal.

The objective, when dealing with SI, should be entirely to think SI, to talk SI, to write SI, to compute SI, to draw SI, to measure SI, and to record data in SI terms. SI is a new language and as such it should be applied to the fullest extent whenever and wherever used. Complete immersion is the best condition for learning.

9.2 NBS Unit Tables

To comprehend SI as a complete "system" it is well to review Tables 9-1, 9-2, 9-3, and 9-4, as four key groupings of SI unit information. Originally published by the National Bureau of Standards, these tabulations provide a progressive, step-by-step introduction to SI.

Essential elements of these tables to be noted are: (a) Quantity Name; (b) Unit Name; (c) Unit Symbol; (d) Expression in terms of SI Base Units and/or other Derived Units. Key characteristics of these tables are:

Table 9-1—emphasizes the seven (7) Base Units of SI;
Table 9-2—lists certain Derived Units in SI each of which has been given a Special Name. These names recognize renowned contributors to the evolution of the process of measurement, such as: newton;
Table 9-3—gives examples of Derived Units expressed entirely in terms of Base Units, such as: m/s;
Table 9-4—shows examples of Derived Units expressed in terms of another Derived Unit with a Special Name and in combination with one or more Base Units, such as: Pa·s.

Table 9-1. SI Base Units

Quantity	SI Base Units Name	Symbol
length	meter	m
mass[1]	kilogram	kg
time	second	s
electric current	ampere	A
thermodynamic temperature[2]	kelvin	K
amount of substance	mole	mol
luminous intensity	candela	cd

BASE UNITS

	SI Supplementary Units	
plane angle	radian	rad
solid angle	steradian	sr

[1] "Weight" in everyday (non-technical) usage generally means "mass."
[2] Wide use is made of "Celsius temperature" (symbol t) defined by $t = T - T_0$ where T is the thermodynamic temperature, expressed in kelvins, and $T_0 = 273.15$ K by definition. The unit "degree Celsius" is thus equal to the unit "kelvin" but the degree Celsius (symbol °C) is a special name used instead of kelvin for expressing Celsius temperature. A temperature interval or a Celsius temperature difference may be expressed in degrees Celsius as well as in kelvins.

9.3 SI Unit Charts

Entirely a coherent system, with all Derived Units on a one-to-one relationship with Base Units and with each other, SI fits readily into a graphical portrayal of unit relationships.

Each of the accompanying charts (Figs. 9-1 to 9-6) is built upon several of the seven Base Units applicable to the particular field of interest being portrayed.

Unit names and symbols are given in approved SI form with unit names in all lower-case letters and unit symbols in roman (upright) type in accordance with international standards. Quantity symbols (and hence, quantity names on this set of charts), by the same international standard, are in italic type. Thus:

Quantity Symbol	Name	SI Units Symbol	Name
L	LENGTH	m	meter
m	MASS	kg	kilogram
F	FORCE	N	newton

These charts are not intended to be all-inclusive but rather to be illustrative of the type of SI Unit Chart which may be constructed by

Ch. 9 CONVERSION TO PREFERRED SI USAGE

Table 9-2. SI Derived Units Given a Special Name[3]

Quantity	S.I. Unit Name	Symbol	Expression in terms of Other Units
frequency (cycles per second)	hertz	Hz	s^{-1}
force	newton	N	$kg \cdot m/s^2$
pressure, stress, modulus	pascal	Pa	N/m^2
energy, work, quantity of heat	joule	J	$N \cdot m$
power, radiant flux	watt	W	J/s
quantity of electricity, electric charge	coulomb	C	$A \cdot s$
electric potential, potential difference, electromotive force	volt	V	W/A
capacitance	farad	F	C/V
electric resistance	ohm	Ω	V/A
conductance	siemens[4]	S	A/V
magnetic flux	weber	Wb	$V \cdot s$
magnetic flux density	tesla	T	Wb/m^2
inductance	henry	H	Wb/A
luminous flux	lumen	lm	$cd \cdot sr$
illuminance	lux	lx	lm/m^2
activity (of a radionuclide)[5]	becquerel	Bq	s^{-1}
absorbed dose, specific energy imparted, kerma, absorbed dose index	gray	Gy	J/kg

DERIVED UNITS GIVEN A SPECIAL NAME

[3] metric ton (symbol t) = 10^3 kg = Mg is acceptable for use *with* SI, although *not* a part of SI.

[4] The siemens was previously called the mho.

[5] Translators' note: this term is more appropriate than the direct translation 'ionizing radiations' of the present French text.

anyone interested in a particular field of technical activity. The pattern of orientation of frequently used Derived Units adopted in these charts will be found to be a serviceable framework on which to build. It has been found advisable not to make any one chart excessively comprehensive since it may become cluttered and lose its main purpose of clarity. Accordingly, a series of several charts, each developed in accordance with a suitable pattern is generally found to be a more useful reference than one overcrowded chart.

Table 9-3. Examples of SI Derived Units, Expressed in Terms of Base Units

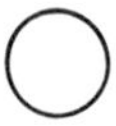

Quantity	SI Unit	Unit Symbol
area[5]	square meter	m^2
volume[6]	cubic meter	m^3
speed, velocity	meter per second	m/s
acceleration	meter per second squared	m/s^2
wave number	one per meter	m^{-1}
density, mass density	kilogram per cubic meter	kg/m^3
current density	ampere per square meter	A/m^2
magnetic field strength	ampere per meter	A/m
concentration (of amount of substance)	mole per cubic meter	mol/m^3
specific volume	cubic meter per kilogram	m^3/kg
luminance	candela per square meter	cd/m^2

DERIVED UNITS EXPRESSED IN TERMS OF BASE UNITS

[5] hectare (symbol ha), 1.0 ha = 10^4 m^2, has been approved for use for land areas in the U.S.A.

[6] liter (symbol L), 1.0 L = dm^3 = 10^{-3} m^3, is approved for use with SI. The international symbol for liter is the lowercase "l", which can be easily confused with the numeral "1". Accordingly, the symbol "L" is recommended for U.S.A. use.

The key to the graphic symbols used in the Chart Series of this book is:

KEY TO CHARTS:

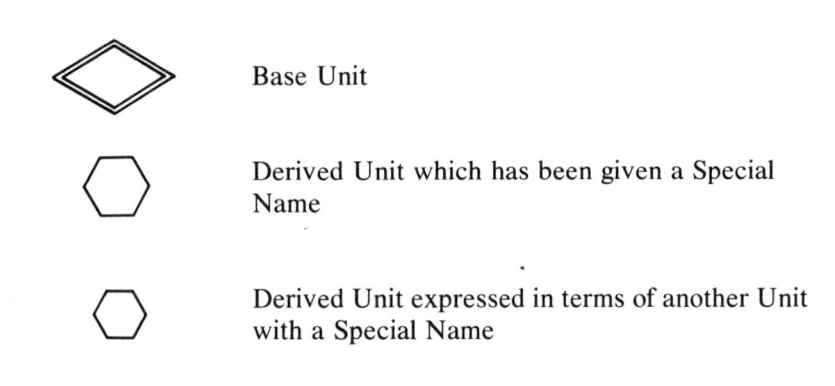

◇ Base Unit

⬡ Derived Unit which has been given a Special Name

⬡ Derived Unit expressed in terms of another Unit with a Special Name

Ch. 9   CONVERSION TO PREFERRED SI USAGE   237

Table 9-4. Examples of SI Derived Units Expressed in Terms of Another Unit with a Special Name

Quantity	Name	Unit Symbol
dynamic viscosity	pascal second	Pa·s
moment of force	newton meter	N·m
surface tension	newton per meter	N/m
power density, heat flux density, irradiance	watt per square meter	W/m^2
heat capacity, entropy	joule per kelvin	J/K
specific heat capacity, specific entropy	joule per kilogram kelvin	J/(kg·K)
specific energy	joule per kilogram	J/kg
thermal conductivity	watt per meter kelvin	W/(m·K)
energy density	joule per cubic meter	J/m^3
electric field strength	volt per meter	V/m
electric charge density	coulomb per cubic meter	C/m^3
electric flux density	coulomb per square meter	C/m^2
permitivity	farad per meter	F/m
permeability	henry per meter	H/m
molar energy	joule per mole	J/mol
molar heat capacity, molar entropy	joule per mole kelvin	J/(mol·K)

DERIVED UNITS EXPRESSED IN TERMS OF ANOTHER UNIT WITH A SPECIAL NAME

⬡   Derived Units Expressed in Terms of Another Unit with a Special Name

◯   Derived Unit expressed in terms of Base Units

(◌)   Other units used with SI

⟶   Unit to the numerator

--⟶   Unit to the denominator

9.4 Numerical Conversion Factors

Inevitably for a transitional period it will be necessary to deal with numerous conversion factors. Successful handling of the process of

238   CONVERSION TO PREFERRED SI USAGE   Ch. 9

MODERNIZED METRIC-SI-System International-A Coherent System of Measuring Units-SI

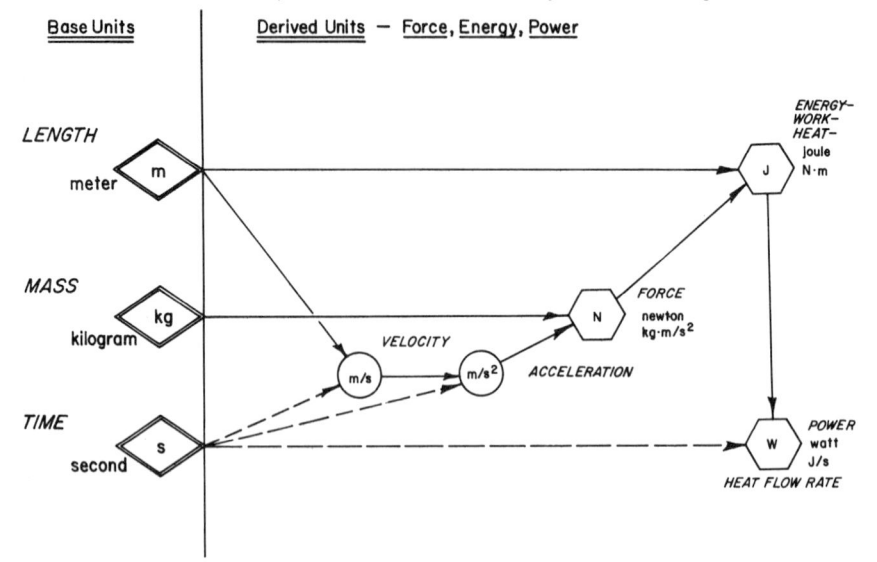

Fig. 9-1. SI Derived Unit Chart—Force, Energy, Power.

MODERNIZED METRIC-SI-System International-A Coherent System of Measuring Units-SI

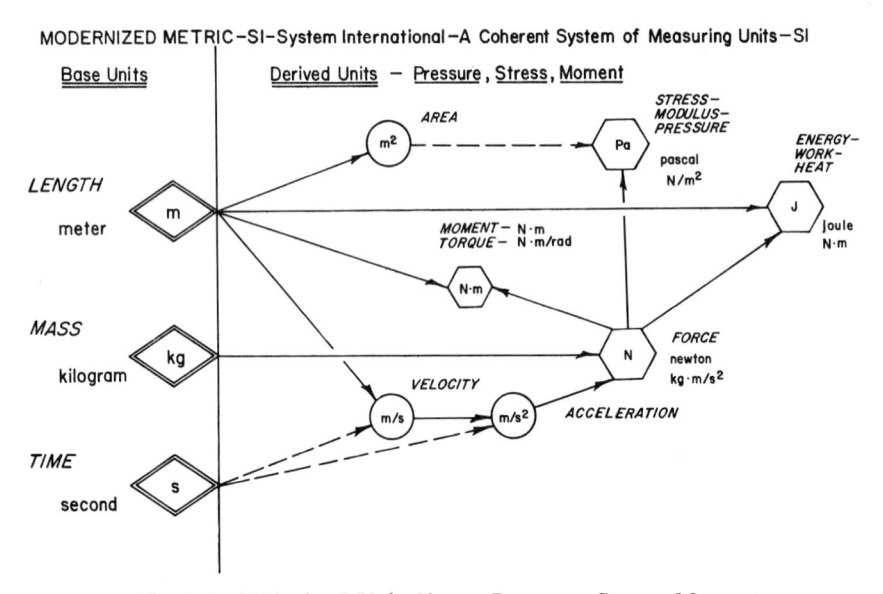

Fig. 9-2. SI Derived Unit Chart—Pressure, Stress, Moment.

Ch. 9   CONVERSION TO PREFERRED SI USAGE   239

MODERNIZED METRIC-SI-System International—A Coherent System of Measuring Units—SI

Base Units | Derived Units — Density, Flow, Temperature, etc (illustrative)

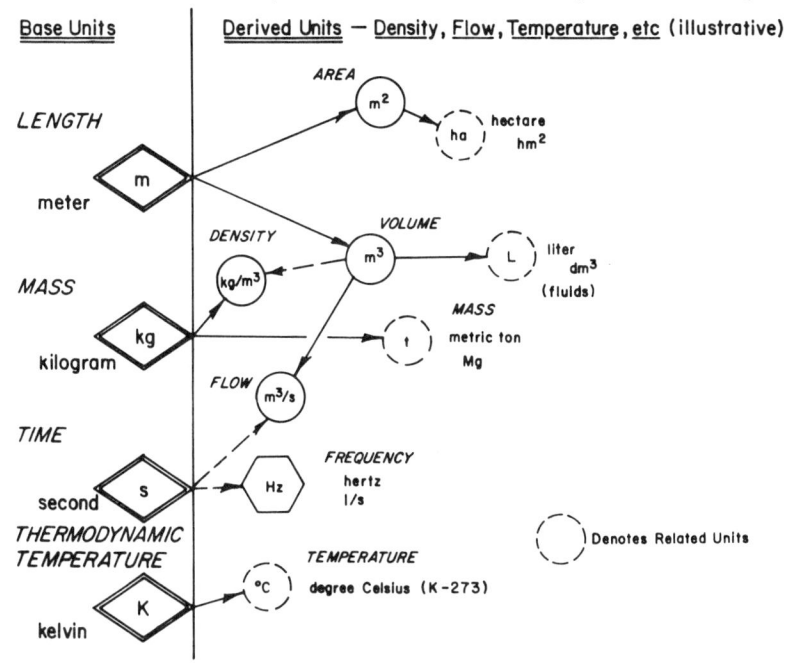

LENGTH — meter (m)
- AREA m² → hectare (ha) hm²
- VOLUME m³ → liter (L) dm³ (fluids)
- DENSITY kg/m³
- MASS t → metric ton Mg

MASS — kilogram (kg)

TIME — second (s)
- FLOW m³/s
- FREQUENCY Hz, hertz, 1/s

THERMODYNAMIC TEMPERATURE — kelvin (K)
- TEMPERATURE °C, degree Celsius (K−273)

( ) Denotes Related Units

Fig. 9-3. SI Derived Unit Chart—Density, Flow, Temperature, etc.

numerical conversion, where needed, will depend on skill in accomplishing the following:

1. Concentrate on *essential categories* such as length, area, volume, mass, force, pressure, stress, moment, torque, energy, work, and power;
2. Select a *few key conversion factors* in each category such that these key factors can be readily related to each other and to the preferred unit multiples in SI;
3. Segregate conversion factors and place *primary* emphasis on "*SI Units* to Customary Units" thus best preparing for the future;
4. Devise means of *checking conversions* by alternate means or by approximations;
5. Keep in mind at all times the matter of *significant digits*, both in the original data and in the conversion result.

Key conversion factors, SI to U.S. Customary units, for various categories of simple quantities are suggested in Table 9-5. This is the

240　　　CONVERSION TO PREFERRED SI USAGE　　　Ch. 9

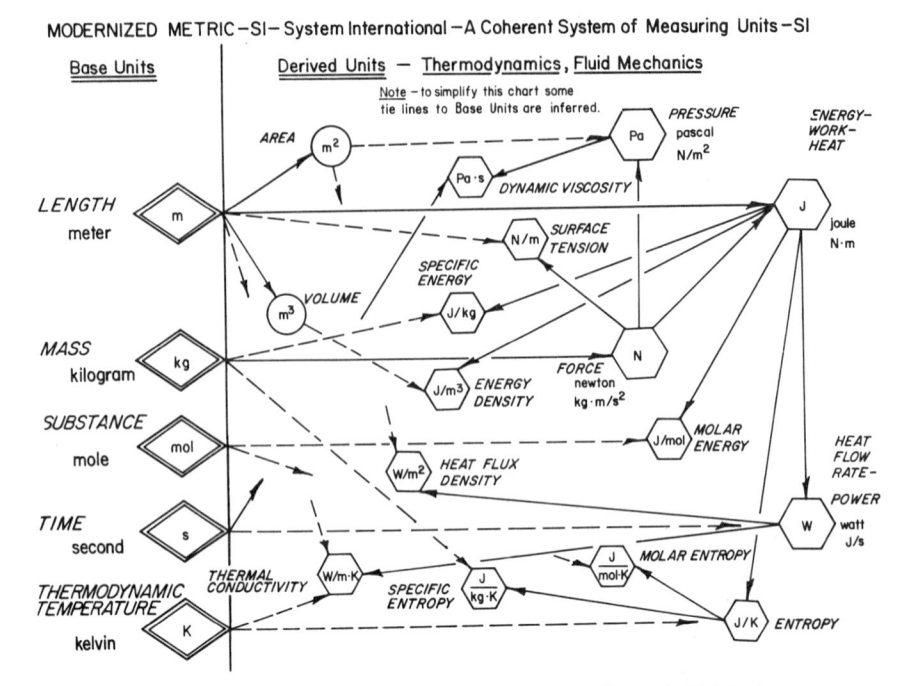

Fig. 9-4. SI Derived Unit Chart—Thermodynamics, Fluid Mechanics.

table on which it is recommended the reader place primary emphasis. Table 9-6 shows the reciprocal conversion factors of the U.S. Customary to SI units for the same quantities. Table 9-7 suggests additional conversion factors for frequently used quantities which involve compound units in each system.

As stated in (4) above, the correctness and appropriateness of any conversion factor *should always be checked by the user*!

9.5 Graphical Conversion

Graphics is the best way to develop a sense of proportion. Accordingly, "conversion" by graphical methods is highly desirable. Since in all cases of the conversion of customary units to SI units there is a simple straight-line relationship, a single ratio factor can be used to construct the SI Chart.

As a first step it is advisable to start with a single everyday unit such as length or distance, and prepare a conversion chart on quadrille ruled

Ch. 9   CONVERSION TO PREFERRED SI USAGE   241

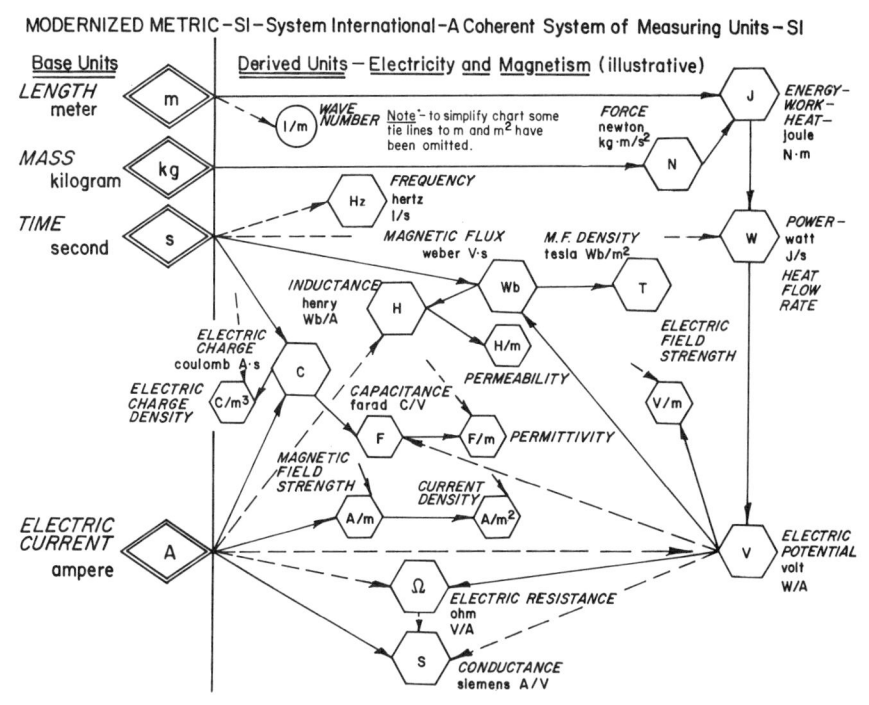

MODERNIZED METRIC—SI—System International—A Coherent System of Measuring Units—SI

Fig. 9-5. SI Derived Unit Chart—Electricity and Magnetism.

paper, as shown in Fig. 9-7. Area, volume, mass, etc., are all good additional examples. Subsequently, a compound unit can be included, such as kilometers per liter to miles per gallon, or Btu per hour to kilojoules per second, etc. On some charts it may be possible, or desirable, to show more than one conversion line under various arrangements as indicated. (See Figs. 9-7 and 9-8.)

It is also highly desirable to prepare such charts on *metric* chart paper, preferably mm. Although this is not necessary it does contribute to a better appreciation of dimensions in mm. If the only chart paper available is in cm, it is advisable to use it in terms of the millimeter unit; i.e., 1 cm = 10 mm.

9.6 Single-Line, Dual-Scale Charts

Conversions may also be shown on a single line, dual-scale, graphical representation similar to that found on many charts, as shown in Fig. 9-9

242   CONVERSION TO PREFERRED SI USAGE   Ch. 9

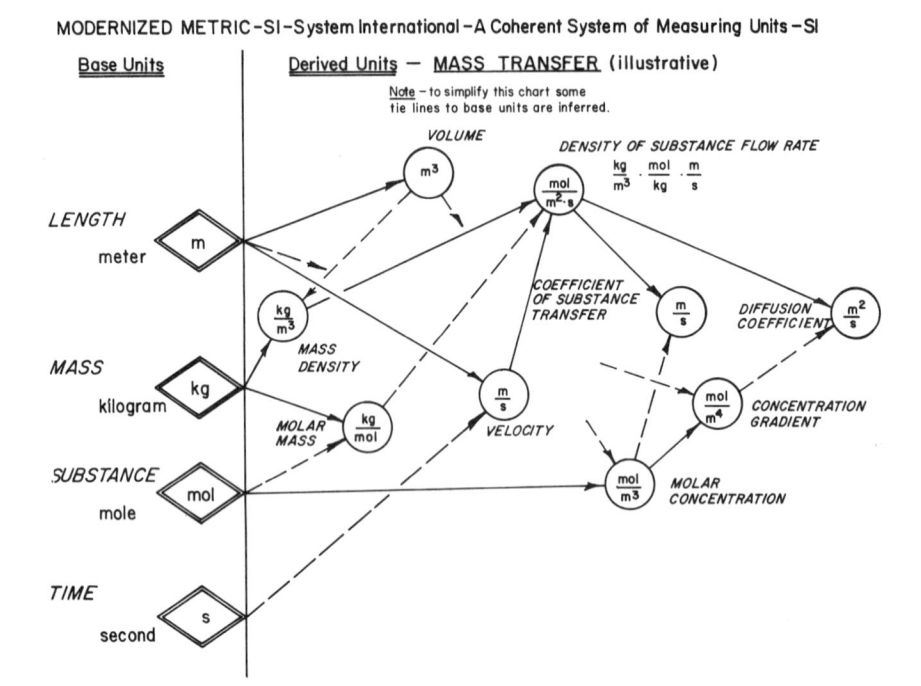

Fig. 9-6. SI Derived Unit Chart—Mass Transfer.

for "Temperature," and in Fig. 9-10 for "Other Quantities." In this case a means of establishing proportionality must be utilized which is really what is being accomplished by plotting on quadrille-ruled paper. Such graphical scales are often found on special slide-rules for conversion.

An important advantage of conversion by graphical representation is that it keeps the user mindful of significant digits. A chart scale should be selected initially on the basis of the significant digits in the original data and the results to be anticipated. No readings beyond those reasonably available by normal interpolation from a chart should therefore be considered as significant.

For a greater degree of precision, any chart of the type in Fig. 9-9 may, of course, be constructed on a larger scale, and may be broken into segments where necessary.

An incident of considerable instructive value may be cited in respect to the relationship of temperature scales. A report of experimental data showed a temperature reading of: $-40°$. The question was raised as to whether this was °C or °F. The response was, "It really doesn't make

Ch. 9　　CONVERSION TO PREFERRED SI USAGE　　243

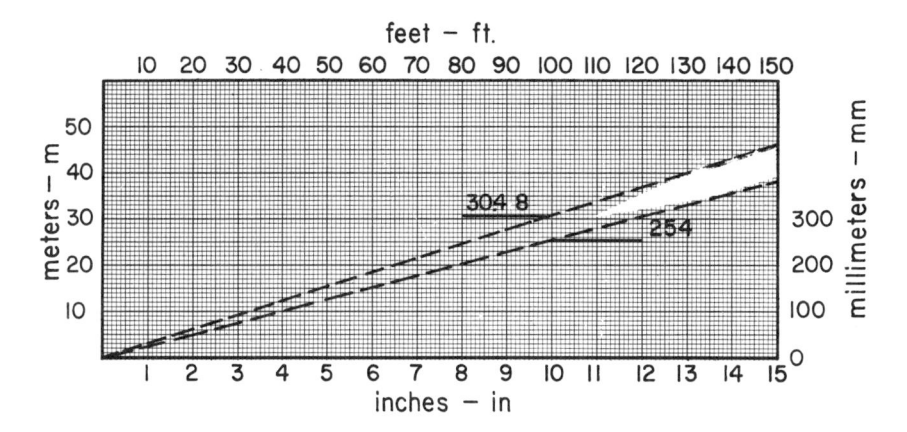

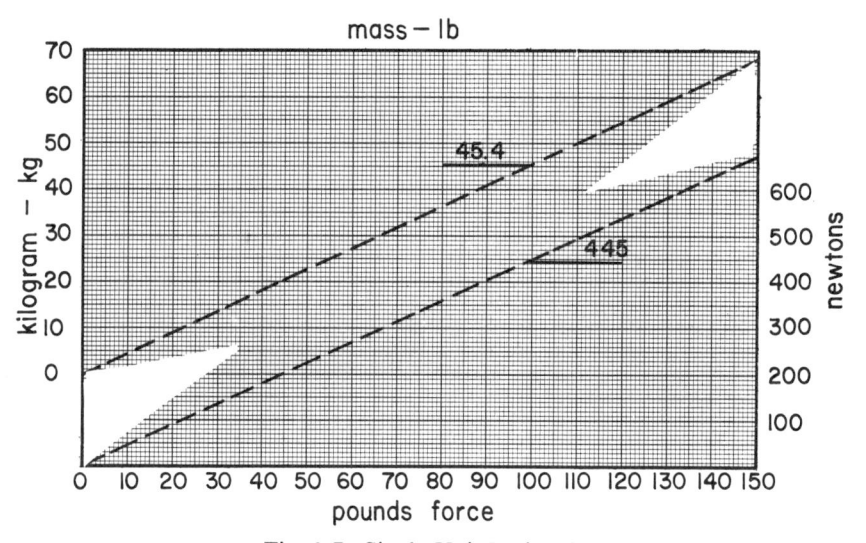

Fig. 9-7. Single Unit Ratio Charts.

any difference"! The response was correct, but only for this *one* unusual instance, as is shown in Fig. 9-9.

9.7 Changeover of Existing Charts—Customary to SI

It is *not* necessary to redraw immediately all charts presently printed in customary units in order to have reference data in SI units.

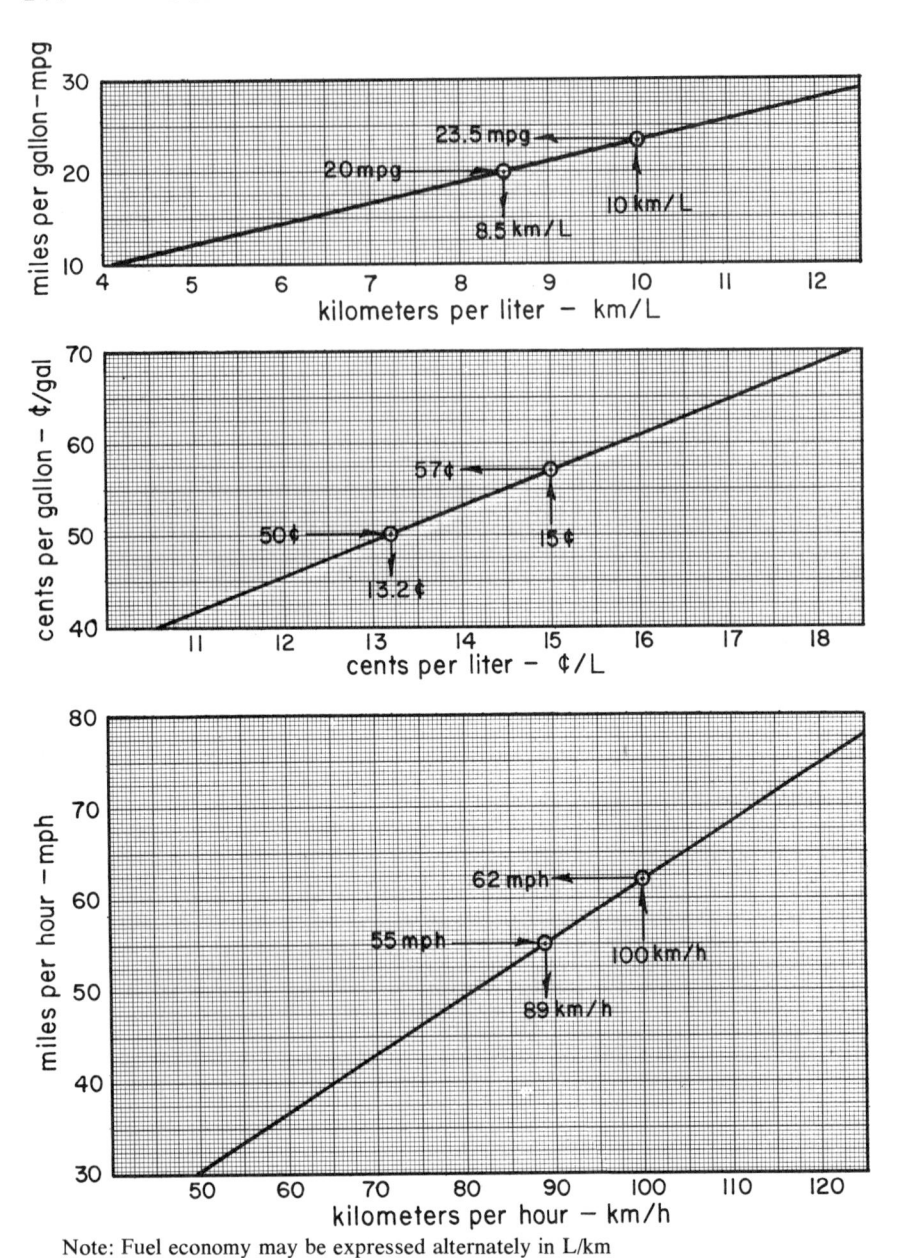

Note: Fuel economy may be expressed alternately in L/km

Fig. 9-8. Compound Unit Ratio Charts (Both fuel consumption and fuel economy may also be stated as: L/100 km.).

Table 9-5. Key Conversion Factors—SI to U.S. Customary Units*

LENGTH	AREA	VOLUME
mm = 0.0394 in.	mm^2 = 0.00155 in.^{2}	mm^3 = 0.00006102 in.^{3}
m = 3.2808 ft	m^2 = 10.7643 ft^2	m^3 = 35.3107 ft^3
m = 1.0936 yd	m^2 = 1.1960 yd^2	m^3 = 1.3080 yd^3
km = 0.6214 mile	km^2 = 0.3861 mi^2	L = 0.2642 gal
	ha=hm^2 = 2.4710 acre	L = 1.0567 qt
MASS	**FORCE**	**PRESSURE-STRESS**
g = 0.0353 oz	N = 3.5970 ozf	kPa = 0.1450 psi
kg = 2.2046 lb	N = 0.2248 lbf	kPa = 0.3346 ft H$_2$O
kg = 0.0685 slug	N = 7.2328 pdl	kPa = 4.0187 in. H$_2$O
Mg = 1.1023 ton	N = 0.1020 kgf	kPa = 7.5008 mm Hg
		MPa = 0.1020 kgf/mm^2
MOMENT	**ENERGY-WORK-HEAT**	**POWER**
N·mm = 0.1416 in.-ozf	J = 0.7376 ft-lbf	W = 0.7376 ft·lbf/s
N·m = 8.8511 in.-lbf	kJ = 0.9478 Btu	W = 3.4122 Btu/hr
N·m = 0.7376 ft-lbf	kJ = 0.2388 kcal	kW = 1.3410 hp
	J = 1.0000 W·s	
	kJ = 0.2778 W·h	

* For more complete information on interpretation and use of these conversion factors, see Tables 8-1 and 9-8 through 9-13.

Many charts have ordinate and abscissa scales shown only on the bottom and left-hand margins. Thus, without redrawing the chart, the comparable SI values can temporarily be shown by supplementary tick or tab marks on the right-hand and top margins, see Fig. 6-5. Another option is to provide alternate or secondary scales.

Various means can be used for determining the proper ratio for constructing the SI scale. The obvious value of "conversion" by chart (as contrasted to conversion by tabulation of data) is that the use of charts promotes and facilitates the development of an essential sense of proportion and of relationship in a visual sense.

Most charts require interpolation in any case and such interpolation can be done almost as readily from improvised SI data points. These SI data points should, of course, be established at integer values (round numbers) as would be the subdivision on any customary chart. Merely converting the existing customary reference points on a chart directly to

Table 9-6. Key Conversion Factors—U.S. Customary to SI Units*

LENGTH	AREA	VOLUME
in. = 25.4000 mm	in.^{2} = 645.2 mm^2	in.^{3} = 16 387 mm^3
ft = 0.3048 m	ft^2 = 0.0929 m^2	ft^3 = 0.0283 m^3
yd = 0.9144 m	yd^2 = 0.8361 m^2	yd^3 = 0.7646 m^3
mile = 1.6093 km	mile^2 = 2.5900 km^2	US gal = 3.7854 L
	acre = 0.4047 ha	US qt = 0.9463 L
MASS	**FORCE**	**PRESSURE-STRESS**
oz = 28.3495 g	ozf = 0.2780 N	psi = 6.8948 kPa
lb = 0.4536 kg	lbf = 4.4482 N	ft H$_2$O = 2.9890 kPa
slug = 14.5939 kg	pdl = 0.1383 N	in. H$_2$O = 0.2488 kPa
ton = 0.9072 Mg	kgf = 9.8067 N	in. Hg = 3.3769 kPa
		mm Hg = 0.1333 kPa
		kgf/mm^2 = 9.8067 MPa
MOMENT	**ENERGY-WORK-HEAT**	**POWER**
in.·ozf = 7.0616 N·mm	ft·lbf = 1.3558 J	ft·lbf/s = 1.3558 W
in.·lbf = 0.1130 N·m	Btu = 1.0551 kJ	Btu/h = 0.2931 W
ft·lbf = 1.3558 N·m	kcal = 4.1868 kJ	hp = 0.7457 kW
	W·s = 1.0000 J	
	W·h = 3.6000 kJ	

* For more complete information on interpretation and use of these conversion factors, see Tables 8-1 and 9-8 through 9-13.

SI would give awkward intermediate decimal values and would be self-defeating in the changeover learning process.

At a convenient date such temporary charts can be redrawn with SI as the main scale. When that time comes it may be advisable to give consideration to showing customary unit equivalents on auxiliary scales, at least for a while, but in such a manner that they can readily be withdrawn in the future.

9.8 Precision of Data

As is well known, the true precision of data is often left vague, especially in the statement of illustrative problems. But in connection with matters of conversion of data from one unit system to another, preciseness becomes an item to which greater attention must be de-

Ch. 9   CONVERSION TO PREFERRED SI USAGE   247

TEMPERATURE

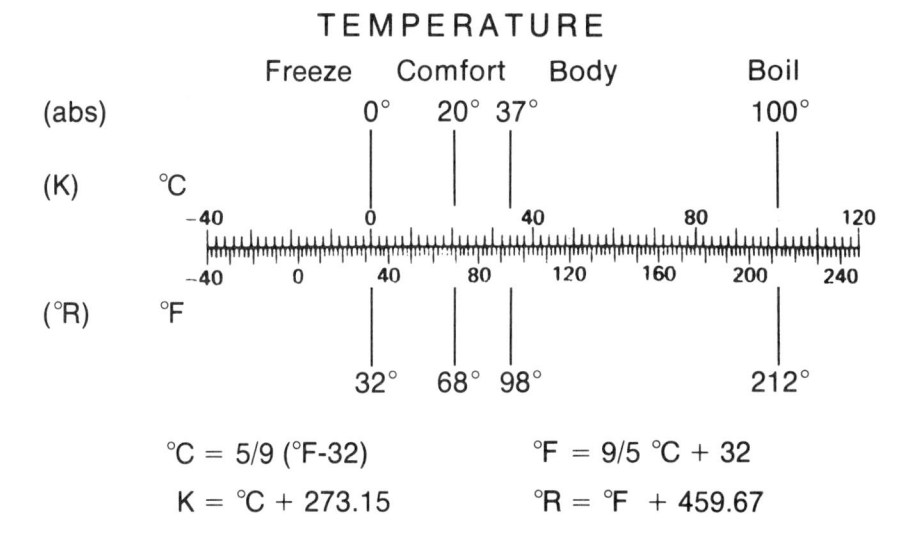

```
                Freeze  Comfort  Body                Boil
(abs)             0°      20°    37°                 100°

(K)      °C
              -40         0         40        80         120
              |—|—|—|—|—|—|—|—|—|—|—|—|—|—|—|—|—|—|—|—|
              -40     0       40      80     120   160    200   240
(°R)     °F
                32°     68°    98°                   212°
```

°C = 5/9 (°F−32)   °F = 9/5 °C + 32

K = °C + 273.15   °R = °F + 459.67

Conversion of Temperature Tolerance Requirements

Tolerance, deg F	±1	±2	±5	±10	±15	±20	±25
Tolerance, K or deg C	±0.5	±1.1	±3	±5.5	±8	±11	±14

Fig. 9-9. Temperature Conversion by Chart.

voted. Strictly speaking, for example:

   5 means 4.5 or *more*, but *less* than 5.5

   5.0 means 4.95 or *more*, but *less* than 5.05

If the quantity described in words as "one mile" is to be restated as 5280 feet, and the latter is to be interpreted as being to the nearest *ten* feet, as would seem to be implied, then the original statement actually should be: 1.00 mile = 5280 feet.

But if the statement is intended to be correct to the nearest *one* foot, then it properly should read:

   1.000 mile = 5280. feet

Such matters are especially important considerations when conver-

248 CONVERSION TO PREFERRED SI USAGE Ch. 9

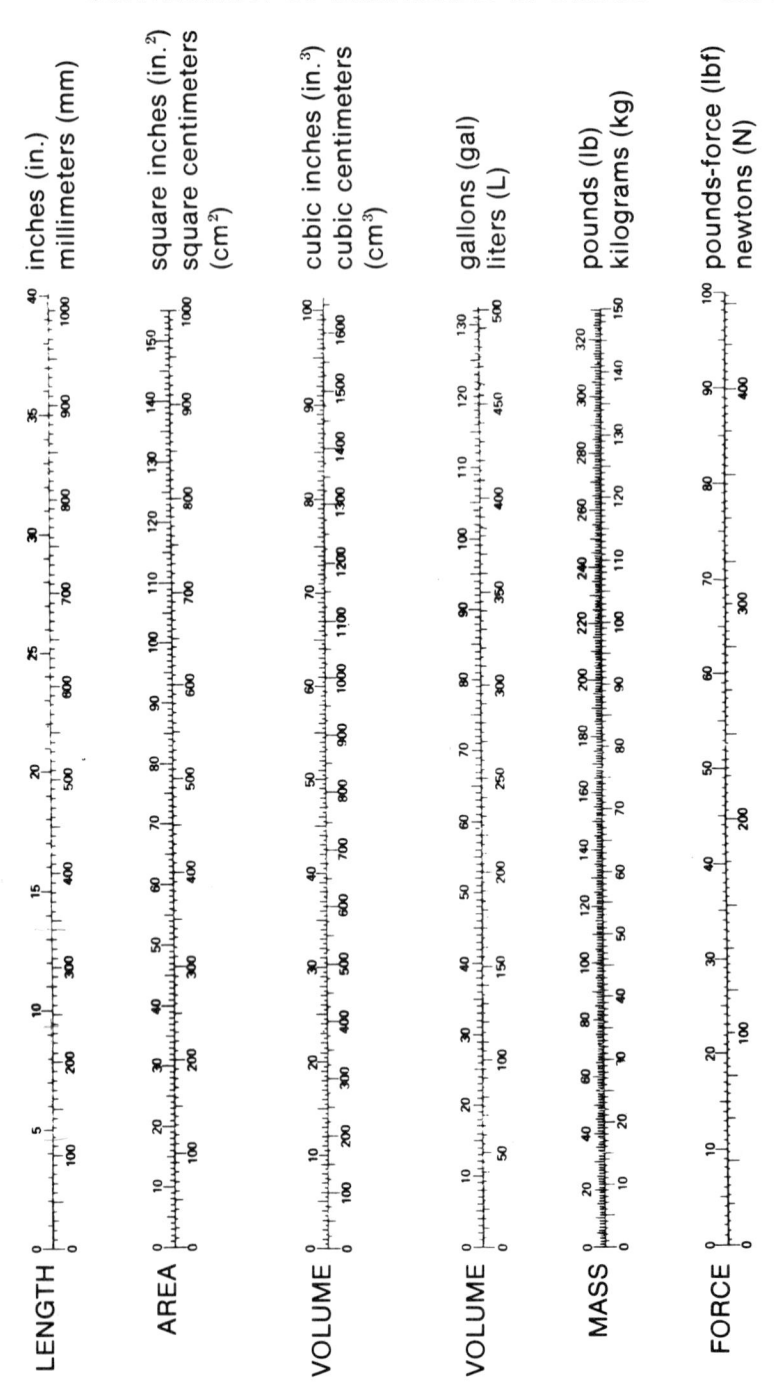

LENGTH: inches (in.) / millimeters (mm)

AREA: square inches (in.²) / square centimeters (cm²)

VOLUME: cubic inches (in.³) / cubic centimeters (cm³)

VOLUME: gallons (gal) / liters (L)

MASS: pounds (lb) / kilograms (kg)

FORCE: pounds-force (lbf) / newtons (N)

Ch. 9   CONVERSION TO PREFERRED SI USAGE   249

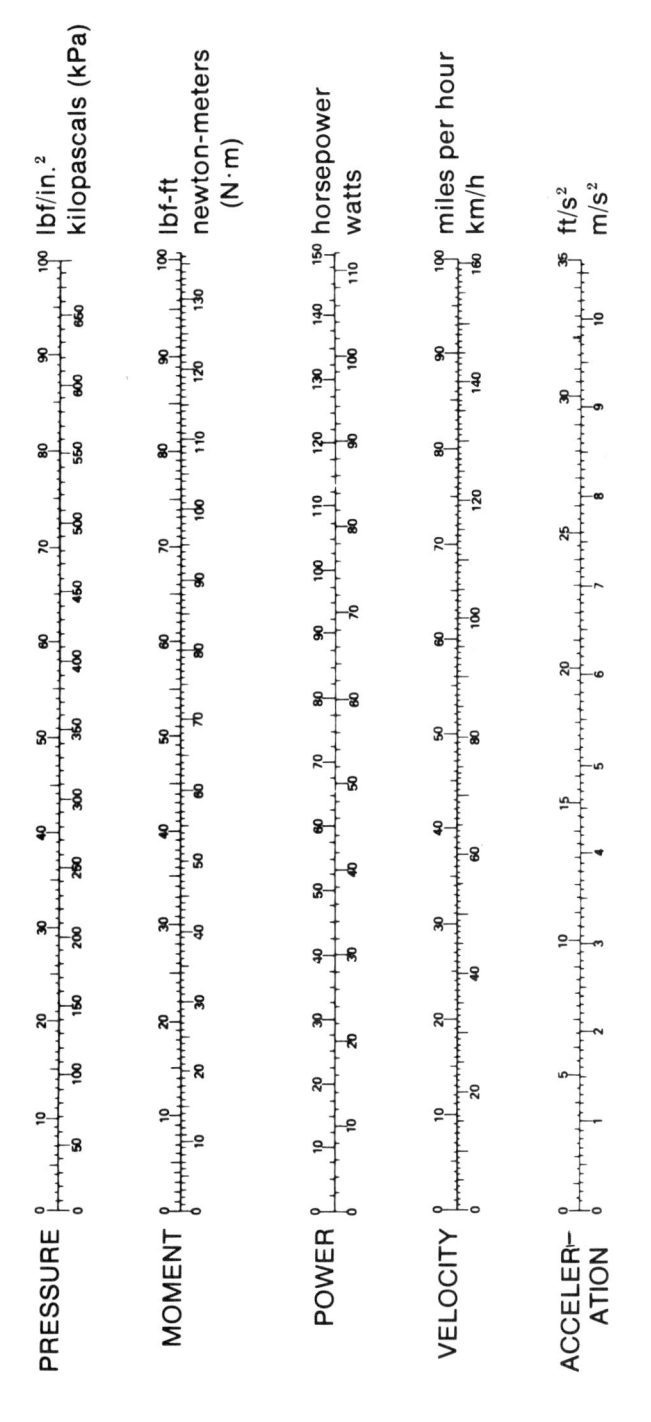

Fig. 9-10. Other Single-line, Dual-scale charts.

Table 9-7. Typical Conversion Factors for Quantities with Compound Units

Q — fluid flow; discharge:	1.0 m³/s = 35.314 ft³/s
	= 22.82 million gal/d
	= 70.05 acre-ft/d
	1.0 L/s = 2.119 ft³/min
	= 15.85 gal/min
e, u, h — specific energy:	1.0 kJ/kg = 0.4299 Btu/lb
c, s, R — specific heat, etc. (mass):	1.0 kJ/(kg·K) = 0.2388 Btu/(lb·°R)
— specific heat, etc. (volume):	1.0 kJ/(m³·K) = 0.1491 Btu/(ft³·°R)
k — thermal conductivity:	1.0 kW/(m·K) = 577.8 Btu/(ft·h·°R)
c, h, U — heat transfer coefficient:	1.0 kW/(m²·K) = 176.1 Btu/(ft²·h·°R)
q'' — heat flux density:	1.0 kW/m² = 317.0 Btu/(ft²·h)
E/v — energy per unit volume:	1.0 kJ/m³ = 0.02684 Btu/ft³
μ — dynamic viscosity:	1.0 kg/(m·s) = 1.0 N·s/m²
	= 1.0 Pa·s = 10 dyne·s/cm²
	= 10 poise = 2419 lb/(ft·h)
ν — kinematic viscosity:	1.0 m²/s = 10⁴ cm²/s = 10 stokes
	= 38 750 ft²/h
I — second moment of area:	1.0 mm⁴ = 2.403 × 10⁻⁶ in.⁴
m/L — mass per unit length:	1.0 kg/m = 0.6720 lb/ft
m/A — mass per unit area:	1.0 kg/m² = 0.2048 lb/ft²
ρ — density (mass per unit volume):	1.0 kg/m³ = 0.06243 lb/ft³

sions to or from SI are to be made, as will be shown in these general illustrations:

Example 9-1. Given the conversion factors:

    miles to kilometers = 1.609   kilometers to miles = 0.6214

By Direct Multiplication	Correct Significant Equivalent
3 mi = 4.827 km	3 mi ≅ 5 km
5 km = 3.107 mi	5 km ≅ 3 mi

also: 3.0 mi ≅ 4.8 km, and, 3.00 mi ≅ 4.83 km, etc.
       5.0 mi ≅ 3.1 mi   5.00 km ≅ 3.11 mi

See additional discussion later under "Rounding of Numbers."

Ch. 9 CONVERSION TO PREFERRED SI USAGE

9.9 Significant Digits

The very best possible practical dictum is to beware of becoming obsessed with conversion factors to the extent that the result is an overabundance of insignificant digits.

It should be kept in mind at all times that the number of digits given for a conversion factor in a reference table reflects only the desire of the author to be of assistance to a wide range of users, any one of whom may have a degree of precision in data which differs markedly from that of another user. Hence, it is the direct responsibility of each user of any conversion table to appraise the precision of his original data and thereby to determine the most suitable level of precision for the statement of his results.

To avoid extracting ridiculous results by indiscriminate use of conversion factors, follow this checklist:

a) Be mindful, at all times, of the definition of the units involved;
b) Keep in mind the precision with which the given data was obtained or is stated;
c) Place an accent mark on the last significant digit and all digits which follow in the data which is being used;
d) In the process of any arithmetic operations (addition-subtraction, multiplication-division) place an accent mark over any digit which results directly from an operation involving another digit with an accent mark;
e) In the resultant quantity, the first digit with an accent mark is the last significant digit in the answer;
f) In general, it should be anticipated that the number of significant digits in the result of a conversion should not exceed by more than one, the *least* number of significant digits that were given in any part of the original data.

Occasionally, of course, it is advisable to carry forward *one* more than the last significant digit through intermediate steps so that its impact may be known and evaluated at the point of final judgment.

The proper interpretation of the number of significant digits present in a stated quantity depends upon two key factors: (a) the number and location of zeroes included in the quantity, and (b) the location of the decimal point. Some examples of correct interpretations are given in the tabulation below:

Quantity	No. of Significant Digits	Quantity	No. of Significant Digits
34$\bar{6}$	3	0.00$\bar{1}$	1
2$\bar{5}$0$\bar{0}$	2	0.10$\bar{0}$	3
250$\bar{1}$	4	0.1$\bar{2}$	2
250$\bar{0}$.	4	3.$\bar{0}$	2
2500.$\bar{0}$	5	20.0$\bar{0}$	4

9.9.1 Addition and Subtraction

Appropriate application of the recommended principles to a typical problem in addition and to other typical problems in subtraction is shown in the following:

Example 9-2

a)   27.2$\bar{4}$ b)   262.$\bar{4}$ c)   262.$\bar{4}$ d)   262.$\bar{4}$
     102.$\bar{6}$ (−) 1$\bar{2}$ (−) 12.$\bar{0}$ (−) 12.0$\bar{3}$
       0.74$\bar{5}$
                         25$\bar{0}$.$\bar{4}$ 250.$\bar{4}$ 250.3$\bar{7}$
     130.$\bar{5}$8$\bar{5}$

Correct statement of result is:

    (a) 130.6   (b) 250   (c) 250.4   (d) 250.4

9.9.2 Multiplication and Division

The recommended principles of "tracking" significant digits are demonstrated in these multiplications:

Example 9-3

       (a)   16 $\bar{3}$4$\bar{0}$ (b)   16 34$\bar{0}$.
                 3.2$\bar{4}$ 3.2$\bar{4}$

             49 0$\bar{2}$$\bar{0}$ 49 0$\bar{2}$$\bar{0}$
           3 2$\bar{6}$8 $\bar{0}$ 3 268 $\bar{0}$
         − $\bar{6}$5$\bar{3}$ $\bar{6}$$\bar{0}$ $\bar{6}$53 $\bar{6}$0

         52 $\bar{9}$$\bar{4}$$\bar{1}$.$\bar{6}$$\bar{0}$ 52 $\bar{9}$$\bar{4}$$\bar{1}$.$\bar{6}$0

Correct Answer: (a) 52 900   (b) 52 900

Ch. 9   CONVERSION TO PREFERRED SI USAGE   253

(c) 16 34$\bar{0}$ (d) 16 34$\bar{0}$
      3.24$\bar{0}$ 0.032$\bar{4}$

      49 02$\bar{0}$ 490 2$\bar{0}$
       3 268 $\bar{0}$ 32 68$\bar{0}$
         653 6$\bar{0}$ $\bar{6}$ $\bar{5}$36 $\bar{0}$
         $\bar{0}\bar{0}$ $\bar{0}\bar{0}\bar{0}$
                                    529.$\bar{4}$$\bar{1}$6 $\bar{0}$
      52 9$\bar{4}$$\bar{1}$.$\bar{6}$$\bar{0}$$\bar{0}$

Correct Answer: (c) 52 940   (d) 529

Comparing (b) to (a), although there are more significant digits in the multiplicand, the number of significant digits in the multiplier still controls and gives the same result. In (c) there are more significant digits in the multiplier and hence, also in the product. In (d) the location of the decimal point does not change the number of significant digits but it does change the statement of the product as compared to (b).

9.10 Rounding of Numbers

Having discussed the matter of "significant digits" and the importance of always considering the "precision of data," it is now essential to examine the process of "rounding" of numbers.

Utmost discretion in rounding is essential since in the changeover from one measuring system to another the units being dealt with in the two systems are frequently of substantially different size. For instance, a quantity that is "rounded" to the nearest meter has an implied precision of ±0.5 m, while a quantity that is "rounded" to the nearest foot has an implied precision of ±0.5 ft. Obviously, the two are quite different. In fact, it will be observed that if a quantity in "feet" to the nearest foot is to be converted to meters, as indicated above, any rounding that is done should be to the nearest (12/39.37) = 0.3 m.

A related example is found in the case of allowable stresses. For instance, structural steel computations in customary units are generally considered to be accurate enough if determined within the nearest 100 psi. Since 1.0 psi is equal to 6.89 kPa, it would be similarly acceptable if stresses in SI units were to the nearest 0.7 MPa. The desired approach (just a bit conservative) would therefore be to give allowable stress to the nearest 0.5 MPa.

Similar consideration would, of course, have to be given to the precision of specified loads, span, section modulus, etc. In summary, the decision needs to be made in such a case as to whether the precision required calls for three or four significant digits. When this is decided, the same criterion should be applied to all factors related to the computation.

Guidance for rounding is embodied in this set of frequently quoted rules:

1. Conversion of quantities should always be done with due regard for appropriate correspondence between the precision of the given data and the precision of the converted data;
2. Do not round either the given data nor the conversion factor before performing the arithmetic operation. Rounding should be applied only to the result;
3. Round the converted quantity to the proper number of significant digits commensurate with its intended precision;

(Text continued on page 260.)

Table 9-8. Length, Area Compilation

To convert from	to	Multiply by
LENGTH		
angstrom	meter (m)	1.000 000* E−10
astronomical unit[13]	meter (m)	1.495 979 E+11
caliber (inch)	meter (m)	2.540 000* E−02
chain	meter (m)	2.011 684 E+01
fathom	meter (m)	1.828 8 E+00
fermi (femtometer)	meter (m)	1.000 000* E−15
foot	meter (m)	3.048 000* E−01
foot (U.S. survey)[12]	meter (m)	3.048 006 E−01
inch	meter (m)	2.540 000* E−02
light year	meter (m)	9.460 55 E+15
microinch	meter (m)	2.540 000* E−08
micron	meter (m)	1.000 000* E−06
mil	meter (m)	2.540 000* E−05
mile (international nautical)	meter (m)	1.852 000* E+03
mile (U.K. nautical)	meter (m)	1.853 184* E+03
mile (international)	meter (m)	1.609 344* E+03

Ch. 9 CONVERSION TO PREFERRED SI USAGE

Table 9-8. Length, Area Compilation (Cont.)

To convert from	to	Multiply by
mile (U.S. statute)[12]	meter (m)	1.609 347 E+03
parsec[13]	meter (m)	3.085 678 E+16
pica (printer's)	meter (m)	4.217 518 E−03
point (printer's)	meter (m)	3.514 598* E−04
rod	meter (m)	5.029 210 E+00
yard	meter (m)	9.144 000* E−01

AREA

acre (U.S. survey)[12]	meter^2 (m^2)	4.046 873 E+03
are	meter^2 (m^2)	1.000 000* E+02
barn	meter^2 (m^2)	1.000 000* E−28
circular mil	meter^2 (m^2)	5.067 075 E−10
ft^2	meter^2 (m^2)	9.290 304* E−02
hectare	meter^2 (m^2)	1.000 000* E+04
in.^{2}	meter^2 (m^2)	6.451 600* E−04
section	meter^2 (m^2)	2.589 998 E+06
township	meter^2 (m^2)	9.323 994 E+07
yd^2	meter^2 (m^2)	8.361 274 E−01

* An asterisk after the sixth decimal place indicates that the conversion factor is exact and that all subsequent digits are zero. All other conversion factors have been rounded to the figures given in accordance with ANS Z210.1-1976 (ASTM E 380-76; IEEE Std. 268-1976), Sec. 4.4. "Rounding Values." Where less than six decimal places are shown, more precision is not warranted.

[12] The 1866 Metric Law defined the ratio of the foot and meter and since 1893 (Mendenhall Order) the U.S. basis of length measurement has been derived from metric standards. In 1959 a refinement was made in the definition of the yard to resolve small differences between this country and abroad, changing its length from 3600/3937 m to 0.9144 m, exactly. The new value is shorter by two parts in a million.

At the same time it was decided that any data in feet derived from and published as a result of geodetic surveys within the U.S. would remain with the old standard (1 ft = 1200/3937 m) until further decision. This foot is named the U.S. survey foot.

Thus all U.S. land measurements in U.S. customary units continue to relate to the meter by the old standard. All the conversion factors for land measure in these tables for units referenced to this footnote are based on the U.S. survey foot, and on the following relationships:

$$1 \text{ rod} = 16\tfrac{1}{2} \text{ feet}$$
$$1 \text{ chain} = 66 \text{ feet}$$
$$1 \text{ mile (U.S. statute)} = 5280 \text{ feet}$$
$$1 \text{ section} = 1 \text{ square mile}$$
$$1 \text{ township} = 36 \text{ square miles}$$

[13] This value conflicts with the value printed in NBS 330 (17). The value requires updating in NBS 330. Also from ASTM E 380-76.

Table 9-9. Volume, Flow Compilation

To convert from	to	Multiply by

VOLUME (Includes CAPACITY)

acre-foot (U.S. survey)[12]	meter³ (m³)	1.233 489 E+03
barrel (oil, 42 gal)	meter³ (m³)	1.589 873 E−01
board foot	meter³ (m³)	2.359 737 E−03
bushel (U.S.)	meter³ (m³)	3.523 907 E−02
cup	meter³ (m³)	2.365 882 E−04
fluid ounce (U.S.)	meter³ (m³)	2.957 353 E−05
ft³	meter³ (m³)	2.831 685 E−02
gallon (Canadian liquid)	meter³ (m³)	4.546 090 E−03
gallon (U.K. liquid)	meter³ (m³)	4.546 092 E−03
gallon (U.S. dry)	meter³ (m³)	4.404 884 E−03
gallon (U.S. liquid)	meter³ (m³)	3.785 412 E−03
gill (U.K.)	meter³ (m³)	1.420 654 E−04
gill (U.S.)	meter³ (m³)	1.182 941 E−04
in.³ [see footnote 16]	meter³ (m³)	1.638 706 E−05
liter [see footnote 17]	meter³ (m³)	1.000 000* E−03
ounce (U.K. fluid)	meter³ (m³)	2.841 307 E−05
ounce (U.S. fluid)	meter³ (m³)	2.957 353 E−05
peck (U.S.)	meter³ (m³)	8.809 768 E−03
pint (U.S. dry)	meter³ (m³)	5.506 105 E−04
pint (U.S. liquid)	meter³ (m³)	4.731 765 E−04
quart (U.S. dry)	meter³ (m³)	1.101 221 E−03
quart (U.S. liquid)	meter³ (m³)	9.463 529 E−04
stere	meter³ (m³)	1.000 000* E+00
tablespoon	meter³ (m³)	1.478 676 E−05
teaspoon	meter³ (m³)	4.928 922 E−06
ton (register)	meter³ (m³)	2.831 685 E+00
yd³	meter³ (m³)	7.645 549 E−01

VOLUME PER UNIT TIME (Includes FLOW)

ft³/min	meter³ per second (m³/s)	4.719 474 E−04
ft³/s	meter³ per second (m³/s)	2.831 685 E−02
in.³/min	meter³ per second (m³/s)	2.731 177 E−07
yd³/min	meter³ per second (m³/s)	1.274 258 E−02
gal (U.S. liquid)/day	meter³ per second (m³/s)	4.381 264 E−08
gal (U.S. liquid)/min	meter³ per second (m³/s)	6.309 020 E−05

* See * footnote of Table 9-10.
[12] See Table 9-8, footnote 12.
[16] The exact conversion factor is 1.638 706 4*E—05. (ASTM E 380-76.)
[17] In 1964 the General Conference on Weights and Measures adopted the name liter as a special name for the cubic decimeter. Prior to this decision the liter differed slightly (previous value, 1.000028 dm³) and in expression of precision volume measurement this fact must be kept in mind. (ASTM E 380-76.)

Ch. 9   CONVERSION TO PREFERRED SI USAGE   257

Table 9-10. Mass, Force Compilation

To convert from	to	Multiply by

MASS

carat (metric)	kilogram (kg)	2.000 000* E−04
grain	kilogram (kg)	6.479 891* E−05
gram	kilogram (kg)	1.000 000* E−03
hundredweight (long)	kilogram (kg)	5.080 235 E+01
hundredweight (short)	kilogram (kg)	4.535 924 E+01
kgf·s^2/m (mass)	kilogram (kg)	9.806 650* E+00
ounce (avoirdupois)	kilogram (kg)	2.834 952 E−02
ounce (troy or apothecary)	kilogram (kg)	3.110 348 E−02
pennyweight	kilogram (kg)	1.555 174 E−03
pound (lb avoirdupois)[18]	kilogram (kg)	4.535 924 E−01
pound (troy or apothecary)	kilogram (kg)	3.732 417 E−01
slug	kilogram (kg)	1.459 390 E+01
ton (assay)	kilogram (kg)	2.916 667 E−02
ton (long, 2240 lb)	kilogram (kg)	1.016 047 E+03
ton (metric)	kilogram (kg)	1.000 000* E+03
ton (short, 2000 lb)	kilogram (kg)	9.071 847 E+02
tonne	kilogram (kg)	1.000 000* E+03

FORCE

dyne	newton (N)	1.000 000* E−05
kilogram-force	newton (N)	9.806 650* E+00
kilopond	newton (N)	9.806 650* E+00
kip (1000 lbf)	newton (N)	4.448 222 E+03
ounce-force	newton (N)	2.780 139 E−01
pound-force (lbf)[19]	newton (N)	4.448 222 E+00
lbf/lb (thrust/weight [mass] ratio)	newton per kilogram (N/kg)	9.806 650 E+00
poundal	newton (N)	1.382 550 E−01
ton-force (2000 lbf)	newton (N)	8.896 444 E+03

* An asterisk after the sixth decimal place indicates that the conversion factor is exact and that all subsequent digits are zero. All other conversion factors have been rounded to the figures given in accordance with ANS Z210.1-1976 (ASTM E 380-76; IEEE Std. 268-1976), Sec. 4.4: "Rounding Values." Where less than six decimal places are shown, more precision is not warranted.

[18] The exact conversion factor is 4.535 923 7*E—01. (ASTM E 380-76.)
[19] The exact conversion factor is 4.448 221 615 260 5*E+00. (ASTM E 380-76.)

Table 9-11. Pressure, Stress, Viscosity Compilation

To convert from	to	Multiply by

PRESSURE OR STRESS (FORCE PER UNIT AREA)

To convert from	to	Multiply by
atmosphere (standard)	pascal (Pa)	1.013 250* E+05
atmosphere (technical = 1 kgf/cm^2)	pascal (Pa)	9.806 650* E+04
bar	pascal (Pa)	1.000 000* E+05
centimeter of mercury (0°C)	pascal (Pa)	1.333 22 E+03
centimeter of water (4°C)	pascal (Pa)	9.806 38 E+01
dyne/cm^2	pascal (Pa)	1.000 000* E−01
foot of water (39.2°F)	pascal (Pa)	2.988 98 E+03
gram-force/cm^2	pascal (Pa)	9.806 650* E+01
inch of mercury (32°F)	pascal (Pa)	3.386 38 E+03
inch of mercury (60°F)	pascal (Pa)	3.376 85 E+03
inch of water (39.2°F)	pascal (Pa)	2.490 82 E+02
inch of water (60°F)	pascal (Pa)	2.488 4 E+02
kgf/cm^2	pascal (Pa)	9.806 650* E+04
kgf/m^2	pascal (Pa)	9.806 650* E+00
kgf/mm^2	pascal (Pa)	9.806 650* E+06
kip/in.^{2} (ksi)	pascal (Pa)	6.894 757 E+06
millibar	pascal (Pa)	1.000 000* E+02
millimeter of mercury (0°C)	pascal (Pa)	1.333 22 E+02
poundal/ft^2	pascal (Pa)	1.488 164 E+00
lbf/ft^2	pascal (Pa)	4.788 026 E+01
lbf/in.^{2} (psi)	pascal (Pa)	6.894 757 E+03
psi	pascal (Pa)	6.894 757 E+03
torr (mm Hg, 0°C)	pascal (Pa)	1.333 22 E+02

VISCOSITY

To convert from	to	Multiply by
centipoise	pascal second (Pa·s)	1.000 000* E−03
centistoke	meter^2 per second (m^2/s)	1.000 000* E−06
ft^2/s	meter^2 per second (m^2/s)	9.290 304* E−02
poise	pascal second (Pa·s)	1.000 000* E−01
poundal·s/ft^2	pascal second (Pa·s)	1.488 164 E+00
lb/ft·h	pascal second (Pa·s)	4.133 789 E−04
lb/ft·s	pascal second (Pa·s)	1.488 164 E+00
lbf·s/ft^2	pascal second (Pa·s)	4.788 026 E+01
rhe	1 per pascal second (1/Pa·s)	1.000 000* E+01
slug/ft·s	pascal second (Pa·s)	4.788 026 E+01
stoke	meter^2 per second (m^2/s)	1.000 000* E−04

* An asterisk after the sixth decimal place indicates that the conversion factor is exact and that all subsequent digits are zero. All other conversion factors have been rounded to the figures given in accordance with ANS Z210.1-1976 (ASTM E 380-76; IEEE Std. 268-1976), Sec. 4.4: "Rounding Values." Where less than six decimal places are shown, more precision is not warranted.

Table 9-12. Energy, Work Compilation

To convert from	to	Multiply by

ENERGY (Includes WORK)

British thermal unit (International Table)[14]	joule (J)	1.055 056	E+03
British thermal unit (mean)	joule (J)	1.055 87	E+03
British thermal unit (thermochemical)	joule (J)	1.054 350	E+03
British thermal unit (39°F)	joule (J)	1.059 67	E+03
British thermal unit (59°F)	joule (J)	1.054 80	E+03
British thermal unit (60°F)	joule (J)	1.054 68	E+03
calorie (International Table)	joule (J)	4.186 800*	E+00
calorie (mean)	joule (J)	4.190 02	E+00
calorie (thermochemical)	joule (J)	4.184 000*	E+00
calorie (15°C)	joule (J)	4.185 80	E+00
calorie (20°C)	joule (J)	4.181 90	E+00
calorie (kilogram, Intern. Table)	joule (J)	4.186 800*	E+03
calorie (kilogram, mean)	joule (J)	4.190 02	E+03
calorie (kilogram, thermochemical)	joule (J)	4.184 000*	E+03
electronvolt	joule (J)	1.602 19	E−19
erg	joule (J)	1.000 000*	E−07
ft·lbf	joule (J)	1.355 818	E+00
ft·poundal	joule (J)	4.214 011	E−02
kilocalorie (International Table)	joule (J)	4.186 800*	E+03
kilocalorie (mean)	joule (J)	4.190 02	E+03
kilocalorie (thermochemical)	joule (J)	4.184 000*	E+03
kW·h	joule (J)	3.600 000*	E+06
therm	joule (J)	1.055 056	E+08
ton (nuclear equivalent of TNT)	joule (J)	4.184	E+09[20]
W·h	joule (J)	3.600 000*	E+03
W·s	joule (J)	1.000 000*	E+00

ENERGY PER UNIT AREA TIME

Btu (thermochem.)/ft²·s	watt per meter² (W/m²)	1.134 893	E+04
Btu (thermochem.)/ft²·min	watt per meter² (W/m²)	1.891 489	E+02
Btu (thermochem.)/ft²·h	watt per meter² (W/m²)	3.152 481	E+00
Btu (thermochem.)/in.²·s	watt per meter² (W/m²)	1.634 246	E+06
cal (thermochem.)/cm²·min	watt per meter² (W/m²)	6.973 333	E+02
erg/cm²·s	watt per meter² (W/m²)	1.000 000*	E−03
W/cm²	watt per meter² (W/m²)	1.000 000*	E+04
W/in.²	watt per meter² (W/m²)	1.550 003	E+03

* See * footnote of Table 9-11.

[14] This value was adopted in 1956. Some of the older International Tables use the value 1.055 04 E+03. The exact conversion factor is 1.055 055 852 62*E+03. (ASTM E 380-76.)

[20] Defined (not measured) value. (ASTM E 380-76.)

(Text continued from page 254.)

4. A quantity stated as a limit must be handled so that the stated limit is not violated;
5. In general, where tolerances are concerned, rounding should be done systematically toward the interior of the tolerance zone so that tolerances are never larger than the original tolerance.

For more detailed discussion of tolerances and true positioning, a manual such as: ASTM E-380, *Metric Practice Guide*, should be consulted.

Another important set of guidelines on "rounding," appropriate to *some* situations, is given in this quotation from ANSI Z-25-1-1940 (R 1961):

> When the digit following the last digit to be retained is 5, the last digit retained should be the closest even value. For example:
>
> a) If the last digit to be retained is an odd number, increase it to the next largest number. Thus 0.1875 will become 0.188.
> b) The last digit to be retained is not changed if it is an even number. Thus 1.0625 becomes 1.062.

9.11 Rounded Values

In making the changeover to SI, critical decisions will be made about new rounded values for many factors widely used in technical work. In some instances the new rounded value in SI may be closer to the actual value than is the present value in customary units. A striking example of this fact is found in the value of the modulus of elasticity of steel, as may be seen from this comparison:

Usually quoted customary value:

$$E = 30 \times 10^6 \, \text{psi}$$

$$E = \frac{30 \times 10^6 (9.8) 144}{2.205 (0.3048)^2} = 206.7 \, \text{GPa}$$

$$\frac{\text{lb}}{\text{in.}^2} \cdot \frac{\text{kg}}{\text{lb}} \cdot \frac{\text{m}}{\text{s}^2} \cdot \frac{\text{ft}^2}{\text{m}^2} \cdot \frac{\text{in.}^2}{\text{ft}^2} = \frac{\text{kg} \cdot \text{m}}{\text{m}^2 \cdot \text{s}^2} = \frac{\text{N}}{\text{m}^2} = \text{Pa}$$

On this basis the most widely used values probably will be:

(1) $E = 200$ GPa $(29.0 \times 10^6 \, \text{psi})$

(2) $E = 205$ GPa $(29.7 \times 10^6 \, \text{psi})$

both of which are closer to the usual actual value for mild steel.

Table 9-13. Heat—Power Compilation

To convert from	to	Multiply by
Heat		
Btu (International Table)·ft/h·ft²·°F (k, thermal conductivity)	watt per meter kelvin (W/m·K)	1.730 735 E+00
Btu (thermochemical)·ft/h·ft²·°F (k, thermal conductivity)	watt per meter kelvin (W/m·K)	1.729 577 E+00
Btu (International Table)·in./h·ft²·°F (k, thermal conductivity)	watt per meter kelvin (W/m·K)	1.442 279 E−01
Btu (thermochemical)·in./h·ft²·°F (k, thermal conductivity)	watt per meter kelvin (W/m·K)	1.441 314 E−01
Btu (International Table)·in./s·ft²·°F (k, thermal conductivity)	watt per meter kelvin (W/m·K)	5.192 204 E+02
Btu (thermochemical)·in./s·ft²·°F (k, thermal conductivity)	watt per meter kelvin (W/m·K)	5.188 732 E+02
Btu (International Table)/ft²	joule per meter² (J/m²)	1.135 653 E+04
Btu (thermochemical)/ft²	joule per meter² (J/m²)	1.134 893 E+04
Btu (International Table)/h·ft²·°F (C, thermal conductance)	watt per meter² kelvin (W/m²·K)	5.678 263 E+00
Btu (thermochemical)/h·ft²·°F (C, thermal conductance)	watt per meter² kelvin (W/m²·K)	5.674 466 E+00
Btu (International Table)/s·ft²·°F	watt per meter² kelvin (W/m²·K)	2.044 175 E+04
Btu (thermochemical)/s·ft²·°F	watt per meter² kelvin (W/m²·K)	2.042 808 E+04
Btu (International Table)/lb	joule per kilogram (J/kg)	2.326 000* E+03
Btu (thermochemical)/lb	joule per kilogram (J/kg)	2.324 444 E+03
Btu (International Table)/lb·°F (c, heat capacity)	joule per kilogram kelvin (J/kg·K)	4.186 800* E+03
Btu (thermochemical)/lb·°F (c, heat capacity)	joule per kilogram kelvin (J/kg·K)	4.184 000 E+03

(Continued on next page)

Table 9-13. Heat—Power Compilation (Cont.)

To convert from	to	Multiply by
	HEAT	
cal (thermochemical)/cm·s·°C	watt per meter kelvin (W/m·K)	4.184 000* E+02
cal (thermochemical)/cm²	joule per meter² (J/m²)	4.184 000* E+04
cal (thermochemical)/cm²·min	watt per meter² (W/m²)	6.973 333 E+02
cal (thermochemical)/cm²·s	watt per meter² (W/m²)	4.184 000* E+04
cal (International Table)/g	joule per kilogram (J/kg)	4.186 800* E+03
cal (thermochemical)/g	joule per kilogram (J/kg)	4.184 000* E+03
cal (International Table)/g·°C	joule per kilogram kelvin (J/kg·K)	4.186 800* E+03
cal (thermochemical)/g·°C	joule per kilogram kelvin (J/kg·K)	4.184 000* E+03
cal (thermochemical)/min	watt (W)	6.973 333 E−02
cal (thermochemical)/s	watt (W)	4.184 000* E+00
clo	kelvin meter² per watt (K·m²/W)	2.003 712 E−01
°F·h·ft²/Btu (International Table) (R, thermal resistance)	kelvin meter² per watt (K·m²/W)	1.761 102 E−01
°F·h·ft²/Btu (thermochemical) (R, thermal resistance)	kelvin meter² per watt (K·m²/W)	1.762 280 E−01
ft²/h (thermal diffusivity)	meter² per second (m²/s)	2.580 640* E−05

POWER

Btu (International Table)/h	watt (W)	2.930 711 E−01
Btu (thermochemical)/h	watt (W)	2.928 751 E−01
Btu (thermochemical)/min	watt (W)	1.757 250 E+01
Btu (thermochemical)/s	watt (W)	1.054 350 E+03
cal (thermochemical)/min	watt (W)	6.973 333 E−02
cal (thermochemical)/s	watt (W)	4.184 000* E+00
erg/s	watt (W)	1.000 000* E−07
ft·lbf/h	watt (W)	3.766 161 E−04
ft·lbf/min	watt (W)	2.259 697 E−02
ft·lbf/s	watt (W)	1.355 818 E+00
horsepower (550 ft·lbf/s)	watt (W)	7.456 999 E+02
horsepower (boiler)	watt (W)	9.809 50 E+03
horsepower (electric)	watt (W)	7.460 000* E+02
horsepower (metric)	watt (W)	7.354 99 E+02
horsepower (water)	watt (W)	7.460 43 E+02
horsepower (U.K.)	watt (W)	7.457 0 E+02
kilocalorie (thermochemical)/min	watt (W)	6.973 333 E+01
kilocalorie (thermochemical)/s	watt (W)	4.184 000* E+03
ton (refrigeration)	watt (W)	3.516 800 E+03

* An asterisk after the sixth decimal place indicates that the conversion factor is exact and that all subsequent digits are zero. All other conversion factors have been rounded to the figures given in accordance with ANS Z210.1-1976 (ASTM E 380-76; IEEE Std. 268-1976), Sec. 4.4: "Rounding Values." Where less than six decimal places are shown, more precision is not warranted.

9.12 Changeover to a Coherent System

One of the most favorable characteristics of SI is that of "coherence." In SI there is no need for duplication of units dealing with the same physical characteristics. This is well-demonstrated by examination of classified lists of units beginning with length and area in Table 9-8. Observe in Table 9-10 how eighteen present units for mass can all conveniently be expressed in terms of the kilogram; twenty-seven present units for volume, as shown in Table 9-9, can all be expressed in terms of the cubic meter (or liter, or milliliter); twenty-four present units for pressure or stress, as given in Table 9-11, can all be expressed in terms of the pascal.

Perhaps the most dramatic simplification occurs in the field of energy, work, and quantity of heat, as shown in Tables 9-12 and 9-13, where a total of twenty-nine units of energy and work, plus twenty-six units of heat can all be expressed in terms of joules. In *all* of these cases resultant or related power may be conveniently expressed in terms of watts (joules per second).

In reviewing the listings in all of these tables, the reader should also keep in mind the wide range of size of coherent units offered by the prefix system, varying, for instance, from μJ to kJ to GJ, depending upon the particular application. Only the decimal point need be shifted and the prefix changed in order to adjust to a variety of magnitudes.

As a matter of convenience, the standard conversion factors are given in these same tables which are reprinted, in part, from ANSI Z 210.1-1976 (ASTM E380-76; IEEE Std 268-1976) with permission.

10

Moving Into the World of SI

10.1 Where to Start

Historically, perhaps the most important single U.S.A. relationship with the metric system occurred in 1866 when an Act of Congress defined the "yard" as equal to 3600/3937 of a meter. Since the "yard" is also defined as "36 inches" (actually, 36.0000 inches) this resulted in the relationship of: 1.000 000 inch = 25.400 05 mm. For many years, for all practical purposes, this was usually stated as: 1 in. = 25.4 mm.

In 1959, the U.S. Department of Commerce defined 0.9144 meter as exactly one yard. The United Kingdom took similar action in 1963. This gives:

$$25.4 \text{ mm (exactly)} = 1.000 \text{ inch.}$$

It will be found highly desirable to use this relationship between millimeters and inches as a datum and to build from that base by a series of simple arithmetic steps, a number of bench marks such as:

12(25.4) = 304.8   304.8 mm = 1.000 ft

3(304.8) = 914.4   914.4 mm = 1.000 yd.

Using the millimeter (mm) as a conceptual basis for building the mental tie-ins is strongly recommended.

The relationship of 25.4 mm equal to one inch serves industry well and is sometimes referred to as the "industrial inch." Cartographers still find a need, however, for the original definition. In 1975 the NBS reaffirmed that the original value, one meter equals 39.37 inches, should continue to be used for all survey operations—geodetic, cadastral, photogrammetric and hydrographic.

Relationships for area and volume, of course, fall out automatically from the square function and the cube function of the above numbers and appear in many tables, including Table 9-5.

For broad technical purposes, the second most important general relationship of SI units to customary units is perhaps best based on the unit mass of water.

An outstanding characteristic of the metric system is the fact that for water, under standard conditions of 4°C (39°F) and 1.0 atmosphere, the mass density $\rho = 1000$ kg/m^3. This results in the following key volume-mass relationships:

$$1.00 \text{ m}^3 \text{ of water} = 1.00 \text{ Mg } (1.00 \text{ metric ton})$$

$$1.00 \text{ dm}^3 \text{ (1.00 L) of water} = 1.00 \text{ kg}$$

From this fact, remembering that the usually quoted customary value for water is 62.4 lb/ft^3, (actually 62.428) the relationship between pounds and kilograms is neatly derived as:

$$\frac{62.43(1.000)}{(12/39.37)^3} = 2205 \frac{\text{lb}}{\text{Mg}} \qquad \frac{\text{lb}}{\text{ft}^3} \cdot \frac{\text{ft}^3}{\text{m}^3} \cdot \frac{\text{m}^3}{\text{Mg}} = \frac{\text{lb}}{\text{Mg}}$$

i.e., 1.000 kg = 2.205 lb. Similarly, it is of value to note that:

$$453.6 \text{ grams} = 1.000 \text{ lb}$$

10.2 Metric is Decimal

As previously stated, one of the prime advantages of the metric system is the decimal characteristic. This is illustrated in Fig. 10-1 where the tie-in of length with mass is also made on the basis of the physical characteristics of water.

Calculations in a decimal system such as metric are greatly simplified since many adjustments in magnitude of quantities are made merely by the relocation of the decimal point. Also, the concept of universal prefixes as a language of decimalization affords a simple and direct means of recording levels of magnitude. The full set of standard prefixes adopted for SI is shown in Table 10-1.

But only a selected few of the prefixes are likely to be incorporated in the preferred usage of any given field of activity. Examples of the selection of prefixes are given under the topic "Preferred Multiples and Sub-Multiples," and will be noted to vary in scope as do the magnitudes encountered in the particular field.

To avoid any hazard of the misplacement of decimal points, it is strongly urged that the selection of preferred prefixes always be such that each interval represents a change in magnitude of 1000. Hence the expressed preference for: mm, m, km; kPa, MPa, GPa, etc.

Ch. 10   MOVING INTO THE WORLD OF SI   267

One cubic meter (m³)

METRIC IS DECIMAL

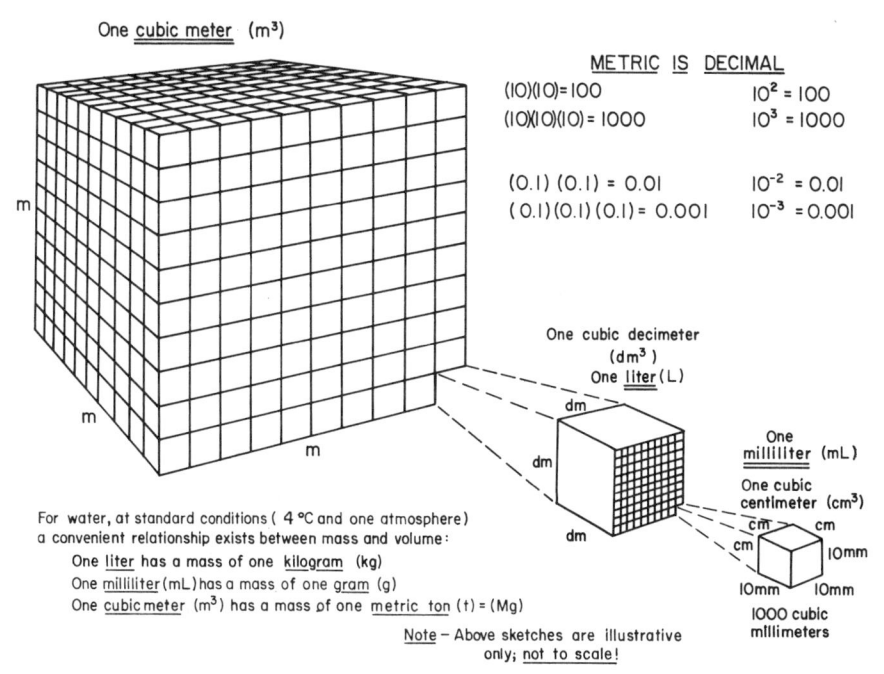

(10)(10) = 100 $10^2 = 100$
(10)(10)(10) = 1000 $10^3 = 1000$

(0.1)(0.1) = 0.01 $10^{-2} = 0.01$
(0.1)(0.1)(0.1) = 0.001 $10^{-3} = 0.001$

One cubic decimeter (dm³)
One liter (L)

One milliliter (mL)
One cubic centimeter (cm³)

1000 cubic millimeters

For water, at standard conditions (4 °C and one atmosphere) a convenient relationship exists between mass and volume:
One liter has a mass of one kilogram (kg)
One milliliter (mL) has a mass of one gram (g)
One cubic meter (m³) has a mass of one metric ton (t) = (Mg)

Note — Above sketches are illustrative only; not to scale!

Fig. 10-1. Metric is Decimal.

Decimalization of all dimensions and data also provides naturally for convenience in the use of calculators and computers of all types.

10.3 Equivalents

One of the essentials in "Going SI" is to develop a "sense of equivalence." At first this is best done in terms of round numbers without getting involved in decimals. A few examples will probably suffice:

300 mm ≈ 1 foot (−)   1 meter ≈ 1 yard (+)
500 g ≈ 1 pound (+)   1 kg ≈ 2 pounds (+)
500 mL ≈ 1 pint (+)
1 liter (L) ≈ 1 quart (+)   4 liters (L) ≈ 1 gallon (+)

Since each of the above is only an approximation (varying 6–10 percent

Table 10-1. SI Unit Prefixes*

Multiplication Factor	Prefix	Symbol	Pronunciation (USA) (1)	Meaning (in USA)	In Other Countries
1 000 000 000 000 000 000 = 10^{18}	exa(2)	E	ex' a (a as in Texas)	One quintillion times (3)	trillion
1 000 000 000 000 000 = 10^{15}	peta(2)	P	as in petal	One quadrillion times (3)	thousand billion
1 000 000 000 000 = 10^{12}	tera	T	as in terrace	One trillion times (3)	billion
1 000 000 000 = 10^{9}	giga	G	jig' a (a as in about)	One billion times (3)	milliard
1 000 000 = 10^{6}	mega	M	as in megaphone	One million times	
1 000 = 10^{3}	kilo	k	as in kilowatt	One thousand times	
100 = 10^{2}	hecto	h(4)	heck' toe	One hundred times	
10 = 10	deka	da(4)	deck' a (a as in about)	Ten times	

Ch. 10 MOVING INTO THE WORLD OF SI

0.1 = 10⁻¹	deci	d(4)	as in decimal	One tenth of
0.01 = 10⁻²	centi	c(4)	as in sentiment	One hundredth of
0.001 = 10⁻³	milli	m	as in military	One thousandth of
0.000 001 = 10⁻⁶	micro	μ(5)	as in microphone	One millionth of
0.000 000 001 = 10⁻⁹	nano	n	nan' oh (an as in ant)	One billionth of (3) — milliardth billionth
0.000 000 000 001 = 10⁻¹²	pico	p	peek' oh	One trillionth of (3) — thousand billionth
0.000 000 000 000 001 = 10⁻¹⁵	femto	f	fem' toe (fem as in femi-nine)	One quadrillionth of (3) — trillionth
0.000 000 000 000 000 001 = 10⁻¹⁸	atto	a	as in anatomy	One quintillionth of (3)

* From "*Metric Editorial Guide*"—American National Metric Council.
(1) The first syllable of every prefix is accented to assure that the prefix will retain its identity.
(2) Approved by the 15th General Conference on Weights and Measures (CGPM), May-June 1975.
(3) These terms should be avoided in technical writing because the denominations above one million are different in most other countries, as indicated in the last column.
(4) While hecto, deka, deci, and centi are SI prefixes, their use should generally be avoided except for the SI unit-multiples for area and volume and nontechnical use of centimeter, as for body and clothing measurement.
(5) Although Rule 1 prescribes upright type, the sloping form is sometimes tolerated in the USA for the Greek letter μ because of the scarcity of the upright style.

from the actual conversions) these equivalencies should *not* be used for anything but general guidance while developing more exact new relationships. These equivalencies do represent some of the key "metric modules" which will be everyday working dimensions of the future.

10.4 Sizes of Familiar Things

The best way to establish bench marks for changeover in unit systems is to develop a sense of equivalence by learning the new "sizes" of familiar things. Preparing a personal reference table drawn from one's everyday experience is an excellent way to begin to do this. An example of such a procedure is found in Table 10-2.

10.5 Classes of Constants

Many frequently used formulas include numerical "constants" which by tradition are sometimes used without a full appreciation of all of the factors they represent. An examination of these constants will indicate that they fall into four classes and should always be recognized as having their origin in one or another of these four classes:

(1) Absolute Numerical Constants which result from the process of developing empirical equations (curve fitting) for the purpose of utilizing experimental data, such as the factors m, b, and n in the expressions:

$$y = mx + b, \quad \text{or,} \quad y = x^n$$

Each of these constants should be investigated as to its origin and the units involved. A slope factor such as "m" is usually a ratio and does not depend on the unit system. An intercept factor such as "b" will depend on the unit used for the ordinate. An exponent such as "n" will generally not change with the unit system as long as it reflects only the power to which the quantity is involved in the physical phenomena being described.

(2) Dimensional Constants which may result from a combination of the given units of physical properties of matter involved and/or a variety of units used for the given conditions. Generally the units used are all in the same unit system. Examples of this class are:

a) A "spring constant" in the expression $t = 2\pi\sqrt{m/k}$ which by analysis of the quantities involved and their usual units, may be

(text continued on page 272)

Ch. 10   MOVING INTO THE WORLD OF SI   271

Table 10-2. Sizes of Familiar Things (Approx.)

Linear Dimension: (approx.)

- 1 mm—0.040 inch
- 1 mm—paperclip wire
- 1 mm—thin dime
- 25 mm—bottle cap diameter
- 25 mm—light bulb base diameter
- 300 mm—width of file folder
- 1 m—baseball bat
- 1 m—normal bounce of a golf ball on a hard surface
- 150 mm by 60 mm—dollar bill (inside border engraving)
- 50 mm by 90 mm by 200 mm— (common brick, incl. joint)

Note—All in group below are in mm

- 1200 × 2400 panel board
- 1500 × 750 desk size; executive
- 1300 × 700 desk size; secretary
- 1200 × 700 desk size; typist
- 700 × 450 office file
- 710 desk height
- 1000 counter height
- 900 office aisle (min.)
- 2000–2100 door height
- 700–800 door width
- 1100 food counter height
- 990–1150 bar height
- 180–186 stair riser
- 240–280 stair tread
- 3000–5500 automobile lengths
- 1500–2100 automobile widths
- 2300 × 4600 parking stall

Mass: (approx.)

- 1 g—paper clip
- 1 g—sweetener packet
- 5 g—5¢ piece
- 50 g—tennis ball
- 50 g—golf ball
- 50 g—medium-size egg
- 7 kg—bowling ball

Volume: (approx.)

- 1 mL—droplets
- 5 mL—teaspoonful
- 15 mL—tablespoonful
- 250 mL—cup
- 1000 mL = 1 liter ≈ 1 quart
- 4 liters(L)—1 gallon
- 20 liters(L)—5 gallons
- 200 liters(L)—commercial drum (55 gallons)
- 400 liters(L)—0.4 m^3 ≈ 100 gallons

Time: (approx.)

- 1 ks ≈ ¼ hour
- 1 Ms ≈ 12 days
- 1 Gs ≈ 32 years

Force:

1.0 N (newton) ≈ a small apple (about 0.1 kg) held on the finger-tips*

1.0 Pa (pascal) ≈ a small apple (about 0.1 kg) exerting a gravitational force uniformly on one square meter*

---

* Since both the newton (N) and the pascal (Pa) are small quantities, they will occur most frequently in terms of: kN, MN; kPa, MPa, GPa, etc.

272 MOVING INTO THE WORLD OF SI Ch. 10

Table 10-2. Sizes of Familiar Things (Approx.) (Cont.)

Pressure:

7 kPa ≈ 1 psi
140 kPa ≈ 20 psi (tire pressure)
140 MPa ≈ 20 000 psi (allow. unit stress in steel)

Velocity (approximate):

1.0 knot ≈ 1.0 nautical mile/hour ≈ 1.9 km/hr ≈ 0.5 m/s
                                    (1.854)   (0.5144)
1.0 km/h ≈ 0.6 mile/hour ≈ 0.9 ft/sec ≈ 0.3 m/s
          (0.6214)   (0.912)   (0.278)
60 km/h ≈ 1.0 km/min ≈ 167 m/s
80 km/h ≈ 50 mph ≈ 1.33 km/min ≈ 222 m/s
1.0 km/liter ≈ 0.4 miles/gallon (0.4250)
1200 km/h ≈ Mach I (speed of sound)

Personal Height and Weight:

An interesting unit comparison is afforded by the following SI interpretation of the recommended optimum height-weight relationship for men: "If your height in centimeters (cm) less your weight (mass) in kilograms (kg) is less than 100, you may need to diet." Now try your age in Gs!

(Continued from page 270.)

evaluated thus:

$$t^2 \propto m/k \quad k \propto m/t^2 \quad m/t^2 \propto \frac{kg}{s^2} = \frac{N}{m} \quad \text{Hence:} \quad k \propto N/m$$

b) A "device constant" such as "K," in the expression $Q = Kh^{1/2}$, used for determining fluid flow through a calibrated meter, may be found as:

$$K \propto Qh^{-1/2} \quad K \propto \frac{m^3}{s} \cdot m^{-1/2} \quad K \propto m^{2.5}/s$$

c) A "special constant" used for limited purposes to express pressure variation with depth in a particular fluid, such as:

$p = k'h$, based on $p = \rho g h$, from which

$$k' = \rho g \quad k' \propto \frac{kg}{m^3} \cdot \frac{m}{s^2} = \frac{N}{m^3} = \frac{Pa}{m} \quad \text{Hence:} \quad k' = Pa/m$$

(3) Unit System Conversion Constants appear when a characteristic such as a friction factor "n" is derived from experimental data recorded in a particular system of units and is later used in the same general formula in another system of units. A classic example of this condition is found in the well-known Manning equation for fluid flow in conduits or open-channels. The respective expressions are:

a) Metric Units (SI):

$$\bar{V} = \frac{1}{n} R^{2/3} S^{1/2}$$

b) Customary Units (ft-lb):

$$\bar{V} = \frac{1.486}{n} R^{2/3} S^{1/2}$$

in which

R = hydraulic radius (area divided by wetted perimeter)
S = slope of hydraulic gradient
$\bar{V}$ = velocity of flow
n = roughness factor.

Since the original work was done and the data recorded in terms of metric units, tabular values of the roughness factor must have dimensional characteristics indicated by this analysis:

$$(R^{2/3} \cdot S^{1/2}) \propto m^{2/3} \cdot (m/m)^{1/2} \propto m^{2/3}$$

but the desired result is:

$$\bar{V} \propto m/s$$

Therefore, the units of the roughness factor must be:

$$n \propto s \cdot m^{-1/3}$$

In the original changeover from metric to English units there was no trouble about the time unit of the second (s) since it is the same in both systems, but it was necessary to introduce the conversion factor of meters to feet to the $1/3$ power. Hence:

$$(3.281)^{1/3} = 1.486$$

The correct statement of this so-called constant, therefore, is:

$$1.486 \ (ft/m)^{1/3}$$

Now as the Manning equation reverts to use with metric units the oft-remembered "constant" of 1.486 becomes unnecessary and does not appear in the formula in SI.

Another example of a unit-sensitive "constant" is found in the specialized formula for the time in seconds for one complete oscillation of a torsional pendulum which is frequently written as:

$$t = 2/3 \sqrt{\frac{\pi W r^2 L}{g d^4 G}}$$

where the customary units are in foot and pound values, except for G which is generally stated in pounds per square inch. The above formula is designed to accept automatically this usual mix of units and sub-units in the customary style.

When it is desired to express the equivalent formula in SI units, the changeover must involve more than the simple substitution of meters for feet and kilograms for pounds. Perhaps the best way to demonstrate this point is to proceed from the more general formula

$$t = 2\pi \sqrt{\bar{I} L / J G}$$

in which $\bar{I} = \frac{1}{2} m r^2$ and $J = \frac{1}{32} \pi d^4$, whence

$$t = 8 \sqrt{\frac{\pi m r^2 L}{d^4 G}}$$

The principal difference in the numerical "constant" 8 versus ⅔ preceding the square root sign, stems, of course, from the fact that in the coherent system of SI, the reference data sources for G state values in the consistent units of N/m^2. Thus the "8" is an absolute number, a true constant.

On the other hand, in the first instance, the factor of 144 has been introduced inconspicuously under the square root sign to adjust for the customary numerial value of G. Thus the factor ⅔ actually embodies units of $(in.^2/ft^2)^{1/2}$.

For this reason basic reference to the most general form of any customary formula is advisable when preparing for any changeover in unit system.

(4) Dimensionless Constants (frequently called "dimensionless numbers") in one sense have a misleading identification since they are obtained from a selected assemblance of quantities with many dimensions.

Actually, such "constants" are more appropriately termed "ratios" and are comprised of quantities selected so as to neutralize dimensional effects. In other words, they represent constants derived from quantities expressed in various units of a particular system but assembled in such a manner that the result is the same numerically regardless of the measuring system used. Such a result is thus dimension free, or independent of the measuring system used, as long as one system is consistently used. This class of "constants" may be said to have a zero order of dimensions, or to be of "neutral dimensions."

Prime examples of this type of constant are the:

Reynolds Number (R) discussed in Section 6.19, and the

Prandtl Number (Pr) discussed in Section 7.12

10.6 Gravitational Attraction

There are some instances in customary technical formulas where the value of gravitational attraction, for all practical purposes, has been considered to be unvarying at the earth's surface. Thus, a standard value of g is sometimes absorbed into an overall constant of a widely used formula. An example of this is given in Section 6.26. Where a changeover of such a formula is to be made the conversion should be made on the basis of one of the following sets of values.

(a) $g = 9.8$ m/s² $= 32.2$ ft/sec² (usually adequate for common usage)

(b) $g = 9.8066$ m/s² $= 32.174$ ft/sec² (for more precise computation)

It will be observed that the use of the simplified value of 9.8 in gravitational computation introduces an error generally of not more than 0.1 percent.

In many formulas the effect of gravitational acceleration appears in the form of $\sqrt{2g}$. In such cases the customary value of 8.02 should be replaced by the SI value of 4.43, or a more precise equivalent value, assuming that g is to be stated in m/s².

10.7 Frequently Used Constants in SI Terms

Each of us meets in everyday work a certain spectrum or array of "constants." Changeover to SI can be facilitated by making a compilation of these with appropriate unit multiples and sub-multiples as given

Table 10-3. Comparison of Fluid Pressure Units—A Case For The Kilopascal*

FIELD OF USE	psi	inch Hg	in. H$_2$O	kPa	MPa	bar	mbar
1. Gasoline engine—air cleaner intake vacuum			2.5 5	0.5 1.0			5 10
2. Hydraulic pump—inlet vacuum		3 10		10 35			100 350
3. Gasoline engine—intake manifold vacuum		5 20		20 70			200 700
4. Power brakes—vacuum tank		15 25		50 80			500 800
5. Atmospheric pressure (absolute)	14.7	29.9	407	101		1.01	1013
6. Fans and centrifugal blowers—static pressure			0.5 5	0.1 1.0			1 10
7. Internal combustion engine—exhaust back-pressure		1 4		3 15			30 150
8. Gasoline engine—fuel pump	1.5 6			10 40		0.1 0.4	
9. Pressurized radiator—pressure relief cap	4 15			30 100		0.3 1	
10. Engine oil pressure	7 30			50 200		0.5 2	
11. City water pressure	30 100			200 700		2 7	
12. LPG—vapor pressure	10 200			70 1400		0.7 14	

13. Pneumatic tire	20 200		140 1400	0.14 1.4	1.4 14
14. Internal combustion engine—cylinder compression	100 200		700 1400	0.7 1.4	7 14
15. Transmission	10 400		70 2800	0.07 2.8	0.7 28
16. Power steering	50 1200		350 8000	0.35 8	3.5 80
17. Hydraulic brakes	200 3000		1 400 21 000	1.4 21	14 210
18. Air Conditioning	0.1 400		0.7 2800		0.007 28
19. Machine tool	80 5000		500 35 000	0.5 35	5 350
20. Hose test pressures	100 20 000		700 140 000	0.7 140	7 1400
21. Air brake	50 150		350 1000	0.35 1.0	3.5 10
22. Farm implement	1200 2800		8 000 19 000	8 19	80 190
23. Industrial hose	1000 5000		7 000 35 000	7 35	70 350
24. Bumper energy absorber gel preload	8500		60 000	60	600

*Source: *Metric Reporter* (ANMC) May 3, 1974. Values in this table are "rounded" and intended to indicate only approximate ranges. For further general interpretation of this table it may be observed that 1.0 psi ≈ 7 kPa, and that 4 inches of water ≈ 1.0 kPa.

in this illustrative listing:

g = standard gravity $\quad$ 9.806 65 m/s² = 32.1740 ft/sec²

θ = angular measure $\quad$ 1.0 rad = 57.2958°

ω = angular velocity $\quad$ 1.0 rad/s = 9.549 r/min

p_0 = standard atmosphere $\quad$ 101.325 kPa = 1.013 25 bar

$\quad$ = 760 mm Hg = 10.33 m H₂O = 1.0332 kgf/cm²

$\quad$ = 29.92 in.Hg = 33.90 ft H₂O = 14.696 lbf/in.²

$\quad$ 1.0 meter of water = 9810 N/m² = 9.81 kPa = 1.42 psi

R_0 = universal gas constant:

8.314 kJ/(kmol·K) = 1.986 Btu/(lb-mol·°R) = 1545 ft·lbf/(lb-mol·°R)

One kmol occupies 22.414 m³ @ 1.00 atmosphere and 0°C. One lb-mol occupies 359.0 ft³ @ 1.00 atmosphere and 32°F.

σ = Stefan-Boltzmann constant:

56.7 × 10⁻¹² kW/(m²·K⁴) = 0.171 × 10⁻⁸ Btu/(ft²·h·°R⁴)

For AIR

R = specific gas constant: $\quad$ 0.2871 kJ/(kg·K) = 0.06856 Btu/
$\quad\quad\quad\quad\quad\quad\quad\quad\quad\quad\quad\quad\quad\quad\quad$ = (lb·°R)
$\quad\quad\quad\quad\quad\quad\quad\quad\quad\quad\quad\quad\quad\quad\quad$ 53.35 ft·lbf/(lb·°R)

c_p = specific heat capacity: $\quad$ 1.005 kJ/(kg·K) = 0.240 Btu/(lb·°R)
(constant pressure)

c_v = specific heat capacity: $\quad$ 0.718 kJ/(kg·K) = 0.1715 Btu/(lb·°R)
(constant volume)

$\gamma = c_p/c_v = 1.40$

M = velocity of sound in air (Mach 1) $\quad$ 344 m/s = 1129 ft/s
(p_0, 20°C, 50% rel.hum.) $\quad\quad\quad\quad\quad\quad$ 1238 km/h = 769.5 miles/hour

10.8 Deciding on Preferred Usage

In order to decide on preferred usage of units and multiples in a particular field it is well to make a survey of the magnitude of quantities to be met, i.e., of the actual range of numbers which will result from the use of the preferred units. An example of an actual survey made in the field of fluid pressure is shown in Table 10-3.

Ch. 10   MOVING INTO THE WORLD OF SI   279

Preparing such a Table automatically develops a new sense of equivalence and also gives an excellent ready reference for changeover.

In general, the objective should be to see that the actual numbers which will result as multipliers with the selected prefix have no greater range than 0.1 to 999.

A similar illustration is given in Table 10-4 which shows how well the kg/m³ serves for identifying ρ, the mass density of materials.

10.9 Series of Preferred Sizes

A once-in-a-lifetime opportunity will occur during many processes of metrication to make new decisions concerning a series of preferred sizes.

Typically, the tendency is to think about increments of size in terms of an arithmetic progression such as: 10, 20, 30, 40, etc., in which each size is larger than its predecessor by a given amount.

But experience in many fields shows that successive size requirements may often be better met if the size series follows a geometric progression

Table 10-4. Mass Densities in SI

The reason for preferring the unit kg/m³ is that it is "coherent," being derived from the SI Base Units for mass and length, and because the numerical values are of convenient size for *all* solids, *all* liquids, and *all* gases at atmospheric pressure, as can be seen from the following examples:

SOLIDS	kg/m³	LIQUIDS	kg/m³	GASES	kg/m³
platinum	21 450	sulfuric acid	1 800	butane	2.412
gold	19 320	battery electrolyte	1 260	propane	1.829
lead	11 340	water	1 000	oxygen	1.330
copper	8 960	mineral oil	900	air	1.206
steel	7 850	alcohol	790	nitrogen	1.165
aluminum	2 700			helium	0.1664
glass	2 600			hydrogen	0.0838
concrete	2 300				
magnesium	1 770				
white oak	770				
white pine	430				
styrofoam	20				

For fluids (both liquids and gases), gram per liter (g/L) is equally acceptable, and will provide the identical numerical values.

Although it is true that ISO-1000 lists six additional units for mass density, the use of such units will only add confusion and the possibility of decimal point errors.

such as: 1, 2, 4, 8, 16, etc., in which each size is larger than its predecessor by a selected percentage or, in other terms, varies in accordance with an exponential relationship.

Combining this view with the decimal characteristics of the metric system, Charles Renard, a French engineer, proposed a variety of size series in which the largest item is ten (10) times the smallest and all intermediate sizes are determined by a common multiple of $\sqrt[n]{10}$, where n is the number of size *intervals* desired. Hence, if in a Renard series there are to be ten size intervals, the ratio of each size in respect to the other will be: $\sqrt[10]{10} = 1.26$. If there are to be five size intervals, the size ratio would be: $\sqrt[5]{10} = 1.58$. Note that this system results in $(n - 1)$ intermediate sizes and accordingly in a total of $(n + 1)$ sizes.

The so-called Renard Numbers result from an adaptation of this concept by ISO along with an agreed-upon rounding for practical

Table 10-5. ISO Recommendation R 3 Series of Preferred Numbers (Renard)

Basic Series			Serial Number*	Calculated Values	Percentage Differences between Basic Series and Calculated Values
R5	R10	R20			
1.00	1.00	1.00	0	1.0000	0
		1.12	2	1.1220	−0.18
	1.25	1.25	4	1.2589	−0.71
		1.40	6	1.4125	−0.88
1.60	1.60	1.60	8	1.5849	+0.95
		1.80	10	1.7783	+1.22
	2.00	2.00	12	1.9953	+0.24
		2.24	14	2.2387	+0.06
2.50	2.50	2.50	16	2.5119	−0.47
		2.80	18	2.8184	−0.65
	3.15	3.15	20	3.1623	−0.39
		3.55	22	3.5481	+0.05
4.00	4.00	4.00	24	3.9811	+0.47
		4.50	26	4.4668	+0.74
	5.00	5.00	28	5.0119	−0.24
		5.60	30	5.6234	−0.42
6.30	6.30	6.30	32	6.3096	−0.15
		7.10	34	7.0795	+0.29
	8.00	8.00	36	7.9433	+0.71
		9.00	38	8.9125	+0.98
10.00	10.00	10.00	40	10.0000	0

* Omitted odd numbers (1, 3, 5, etc.) occur in R 40 series not shown.

Ch. 10   MOVING INTO THE WORLD OF SI   281

purposes. The basis of the R5, R10, and R20 preferred-size series is shown in Table 10-5. Note the consistently small percentage difference between the calculated Renard values and the basic ISO series. Also note that the R40 series can readily be derived from this table, if desired.

These series give sizes which, in general, are doubled every three terms in the R10 series, every six terms in the R20 series, and every twelve terms in the R40 series.

These preferred intervals may be suitably used in the development of size series for products involving length, area, volume, mass, power, etc. The concept may, of course, be applied indefinitely by extrapolation through multiples and sub-multiples of 10.

10.10 Paper Sizes in Metric

The international A-Series of paper sizes is frequently referred to in recommendations for written material and drawings.

The basic size A0 is derived from a surface area of one square meter

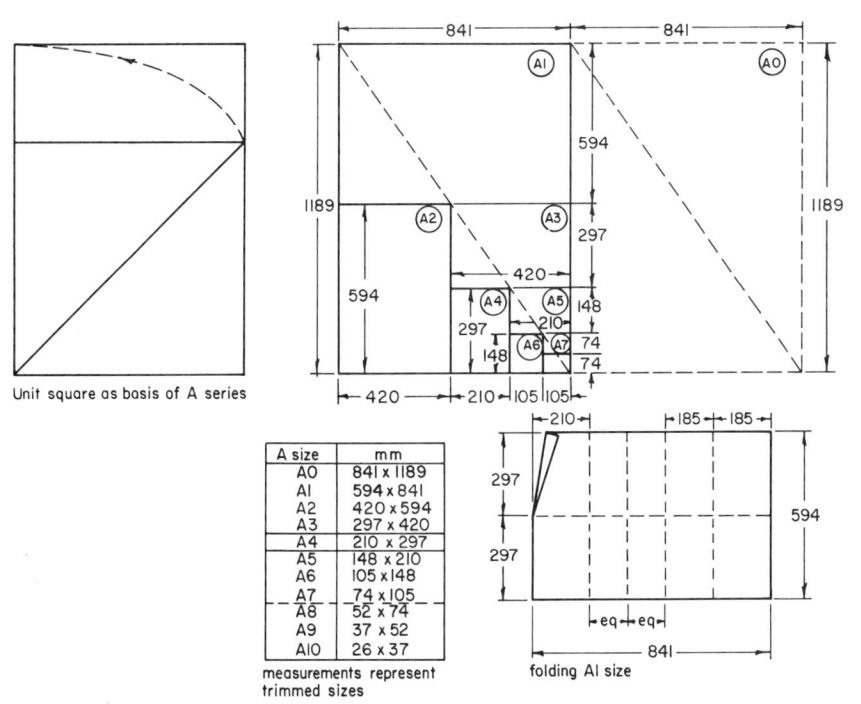

A size	mm
A0	841 x 1189
A1	594 x 841
A2	420 x 594
A3	297 x 420
A4	210 x 297
A5	148 x 210
A6	105 x 148
A7	74 x 105
A8	52 x 74
A9	37 x 52
A10	26 x 37

measurements represent trimmed sizes

Fig. 10-2. Metric paper size series.

in a rectangle of proportions $1:\sqrt{2}$. The resultant full size is 841 × 1189 mm. Smaller sizes are obtained by halving the larger dimension each time. Larger sizes are obtained by doubling the shorter dimension.

Nearest to the customary letter-size sheet of $8\frac{1}{2}$ × 11 inches is the A4 (210 × 297) which is slightly narrower and a little longer.

Closest to large-size customary drawing paper is the A1 (594 × 841) which may be considered the near equivalent of 24 × 36 inches (actual size is 23.4 × 33.1 inches).

In 1975 the American Paper Institute (API) recommended against abrupt size changes and suggested that for the time being there should be no change in actual paper sizes. These two steps were proposed:

1. Paper sizes should be stated in millimeters (mm) using rounded values. Thus, $8\frac{1}{2}$ by 11 becomes 216 by 279.
2. The basis weight should be stated in grams per square meter (g/m^2) rounded. For example, 20-pound paper (17" by 22", 500) equals 75.2 grams per square meter, say, 75 g/m^2.

Note that this is a decision for "soft conversion" at this point in changeover.

Since the main reason for adoption of the metric system is simplicity and potential for modularity, there is good reason to believe that future paper sizes may well become:

New Size—mm	Comparable Size—inches
800 × 1200	
600 × 800	24 × 36 and 24 × 30
400 × 600	
200 × 300	$8\frac{1}{2}$ × 11 and $8\frac{1}{2}$ × 14
150 × 200	

10.11 Drafting Policy

Many actual case-history references are available for guidance on the determination of advisable adjustments in drafting policy for metrication. An overall view of these experiences suggests these guidelines:

(1) Plan to take advantage of the fact that an appreciable saving of time results for both the draftsman and the user when metric drawings can be marked: "All dimensions in millimeters (mm)," or, "All dimensions in meters," as the case may be.

(2) Dual dimensioning on the drawing itself is to be avoided if at all possible, using one of these plans:

   a) Go directly to all metric drawings;
   b) If necessary, add a table of equivalent dimensions on the same sheet, giving SI as prime reference; this can frequently be done with the aid of a computer print-out for paste-on and should be done in a manner that will permit easy removal at a later date.

(3) For a relatively simple drawing give dimensions in terms of code numbers or letters and provide separate key tables, the prime one for SI and a supplementary one for customary units—the latter to be available only as long as needed.

10.12 Drafting Scales

Commonly used drafting scales will change somewhat but all scales will, of course, be on the same decimal basis. There will no longer need to be any difference in concept between an architect's scale and an engineer's scale. Comparisons between present customary scales and recommended ISO scales are shown in Table 10-6.

Table 10-6. Scales for Drawings

Customary Scale	Comparable ISO Scale	SI Relationship
Site Plans		
$\frac{1}{16}'' = 1'0''$	1:200	5 mm = 1 m
$\frac{1}{8}'' = 1'0''$	1:100	10 mm = 1 m
$1'' = 20'$	1:200	5 mm = 1 m
$1'' = 50'$	1:500	2 mm = 1 m
$1'' = 100'$	1:1000	1 mm = 1 m
Building Designs		
$\frac{1}{8}'' = 1'$	1:100	10 mm = 1 m
$\frac{1}{4}'' = 1'$	1:50	20 mm = 1 m
Details		
$\frac{1}{2}'' = 1'0''$	1:20	50 mm = 1 m
$\frac{3}{4}'' = 1'0''$	1:10	100 mm = 1 m
$1'' = 1'0''$		
$1\text{-}\frac{1}{2}'' = 1'0''$	1:10	100 mm = 1 m
$3'' = 1'0''$	1:5	200 mm = 1 m
$1' = 1'$ (full size)	1:1 (full size)	

284 MOVING INTO THE WORLD OF SI Ch. 10

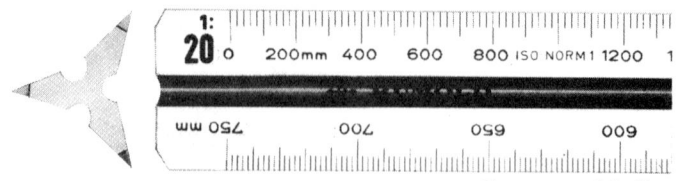

Courtesy of J. Rumold K G

ISO Norm 1 includes 1:20 and 1:2.5 (shown above), also 1:1 and 1:10, as well as 1:5 and 1:50. ISO Norm 2 (not shown) includes 1:200 and 1:2500, also 1:100 and 1:1000, as well as 1:500 and 1:10560. Other scales are available as shown below.

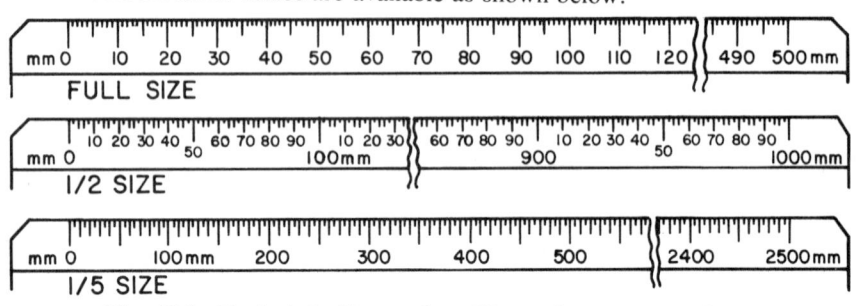

Fig. 10-3. Typical drafting scales. (Illustrative; not to scale.)

SI scales will be either 300 mm or 500 mm in actual graduated length but the graduation pattern will vary in accordance with ISO standards and will be similar to those shown in Fig. 10-3.

10.13 Surveying

Indications are that the changeover to metric in land and engineering surveys will proceed along the following lines:

(a) Distances will be measured in terms of the meter and multiples thereof. Where essential, or advisable, the equivalent value in feet will be given in parentheses;
(b) The surveyor's "foot" will continue to be based on the official U.S. definition of a "yard" as equal to 3600/3937 of a meter;
(c) Measuring tapes will be 30-, 50-, and 100-m lengths, with various styles of graduation including meters, decimeters, centimeters, and millimeters, as may be suitable, but all readings from such

tapes will be directly in decimal parts of the meter, thus: 76.392 m, etc.;
(d) Land area computations and recordings will show values in m² or in ha, depending on size of parcel. The km² is to be reserved for geographical or statistical descriptions of large land areas;
(e) Angles, bearings, and azimuths will continue to be stated in the sexagesimal system of degrees, minutes, and seconds;
(f) Elevations of bench marks and other reference points will be stated in terms of meters;
(g) Standard level rods will move to two-meter length (plus extension) with graduations in multiples of millimeters;
(h) Contours will be shown at 0.5-, 1.0-, 2.0-, and 5.0-m intervals;
(i) Station and offset values will be in terms of the meter, thus Station 1 + 00 being 100 m from Station 0 + 00. Sub-stations and offsets will be in terms of meters;
(j) State plane coordinates will be listed and shown in meters;
(k) Metric graphical scales will appear on all plots and drawings;
(l) Property deeds and legal descriptions will be converted to the metric system only if and when conveyance or subdivision occurs.

10.14 Construction Modules and Sizes

The most significant opportunity for the construction industries at the time of metrication will be the chance to advance dimensional coordination and modular construction.

Metrication does not depend in any way upon dimensional coordination, but as the sizes of construction elements are reviewed there is automatically presented a vast opportunity for rationalization and coordination.

The 100-mm metric module and the 300-mm multi-module, with the many integer sub-modules, present unlimited new opportunities for imaginative and innovative decisions in construction.

It is now clear that the standard wall-panel size of the future will be 1200 × 2400 mm (hard-conversion) rationalized dimensions. The actual dimensions of lumber sizes (studs, joist, planks, beam, and stringers) will not be changed but a soft conversion will occur and these will be described in millimeter designations.

The customary nominal sizes and the customary "board foot" will gradually become obsolete.

As has been mentioned earlier, under "Strength of Materials," there is most likely to be "soft-conversion" of rolled steel structural shapes such as wide-flange sections. That is, depths and properties will not be changed (with a few exceptions for reasons other than metrication) but the dimensions and properties will be described in terms of millimeters. On the other hand, other structural shapes such as angles and zee-bars will probably be hard-converted, i.e., the leg dimensions and thicknesses will actually be adjusted to convenient millimeter values.

Iron pipe sizes and screw-fittings are not apt to be modified, but plastic pipe, tubing, and fittings will undoubtedly become increasingly available in metric sizes.

10.15 Building Code Changeover to SI

I—*LINEAR DIMENSIONS*

(a) It is generally recommended that in the review of building codes and specifications for changeover to the International System of Measuring Units (SI), or the development of alternate codes in SI, *first preference* be given to the substitution for foot and inch dimensions, the following preferred metric dimensions, and derivatives thereof:

(1) Customary	SI	(2) Customary	SI	(3) Customary	SI
ft	mm	inches	mm	inch	mm
1.0	300	12	300	1	25
2.0	600	10	250	$\frac{7}{8}$	21
3.0	900	9	225	$\frac{3}{4}$	18
4.0	1200	8	200	$\frac{5}{8}$	15
8.0	2400	6	150	$\frac{1}{2}$	12
10.0	3000	4	100	$\frac{3}{8}$	9
12.0	3600	3	75	$\frac{1}{4}$	6
20.0	6000	2	50	$\frac{1}{8}$	3
		1	25		

(b) In applying these recommendations it should be recognized that the above table represents a reduction of 1.6% in each dimension of Cols. 1 and 2, and a reduction of 6% in each dimension of Col. 3 except in that for one inch.

(c) Where such reductions can not be generally accepted a study of acceptable tolerances should be conducted and a specific alternate recommendation made.

II—*OTHER DIMENSIONS* Approximate equivalents for other frequently encountered units in construction are:

	Customary Units	SI Units		Customary Units	SI Units
Area:	1.0 sq in.	600 mm²	*Pressure, Stress:*	1.0 psi	7 kPa
	1.0 sq ft	0.1 m²	*Loading:*	100 psf	5.0 kPa
	1.0 sq yd	0.8 m²	*Flow:*	1.0 mgd	0.044 m³/s
	1.0 sq mi	2.6 km²		1.0 cfs	0.028 m³/s
	1.0 acre	0.4 hm²		1.0 acre-ft/d	0.014 m³/s
Volume:	1.0 cu in.	16 000 mm³		1.0 cfm	0.5 L/s
	1.0 cu ft	0.3 m³		1.0 gpm	0.06 L/s
	1.0 cu yd	0.75 m³	*Energy:*	1.0 Btu	1.0 kJ
	1.0 gal	4.0 L		1.0 hp	0.75 kW
	1.0 qt	1.0 L			

Where more precise changeover is believed necessary the customary tolerances and experiences relating thereto should be checked and justified before more exact conversions are recommended, with new SI tolerances.

10.16 Road Map

Many individuals planning to embark on a new venture refer colloquially to the advisability of preparing a "road map." This can well be done in SI and be made to have some real pay-off in terms of better acquaintanceship with that well-known unit of distance, the kilometer (km), see Fig. 10-4.

10.17 A Global Evolutionary Movement

Perhaps of greatest importance is the need to widely develop an understanding that SI did not just happen. SI is *not* just the favorite idea of a few persons. The impact of SI is *not* on only one part of the world.

SI is the most recent major step in an orderly, positive, and continuing evolutionary process in the world of measurements. SI represents a

288 MOVING INTO THE WORLD OF SI Ch. 10

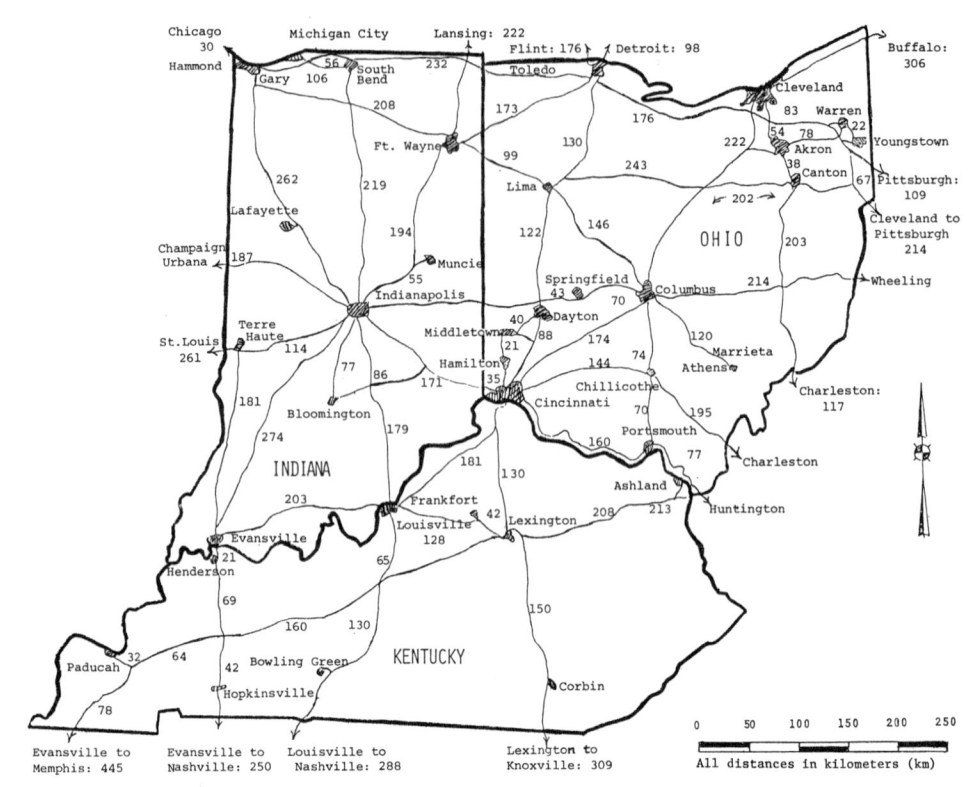

Fig. 10-4. SI "road map."

movement of high potential for developing the first really universal language for all mankind.

An excellent perspective on the metrological background of SI and the mechanisms for making SI grow and flourish beneficially, is given in Appendices I and II. These succinct and impressive historical sketches are highly recommended for additional reading.

Appendix I[1]

Organs of the Metre Convention—BIPM, CIPM, CGPM

The *International Bureau of Weights and Measures* (BIPM) was set up by the *Metre Convention* signed in Paris on 20 May, 1875, by seventeen States during the final session of the Diplomatic Conference of the Metre. This Convention was amended in 1921.

BIPM has its headquarters near Paris, in the grounds 43 520 m^2 of the Pavillon de Breteuil (Parc de Saint-Cloud), placed at its disposal by the French Government; its upkeep is financed jointly by the Member States of the Metre Convention.*

The task of BIPM is to ensure worldwide unification of physical measurements; it is responsible for:

—establishing the fundamental standards and scales for measurement of the principal physical quantities and maintaining the international prototypes;
—carrying out comparisons of national and international standards;
—ensuring the co-ordination of corresponding measuring techniques;
—carrying out and co-ordinating the determinations relating to the fundamental physical constants.

BIPM operates under the exclusive supervision of the *International Committee for Weights and Measures* (CIPM), which itself comes under the authority of the *General Conference on Weights and Measures* (CGPM).

---

* As at 31 December 1976, forty-four States were members of this Convention:
Argentina (Rep. of), Australia, Austria, Brazil, Belgium, Bulgaria, Cameroon, Canada, Chile, Czechoslovakia, Denmark, Dominican Republic, Egypt, Finland, France, German Democratic Rep., Germany (Federal Rep. of), Hungary, India, Indonesia, Iran, Ireland, Italy, Japan, Korea, Mexico, the Netherlands, Norway, Pakistan, Poland, Portugal, Rumania, Spain, South Africa, Sweden, Switzerland, Thailand, Turkey, U.S.S.R., United Kingdom, U.S.A., Uruguay, Venezuela, Yugoslavia.

[1] Reprinted from Special Publication 330-1977, *The International System of Units (SI)*, U.S. Department of Commerce, National Bureau of Standards.

The General Conference consists of delegates from all the Member States of the Metre Convention and meets at least once every six years. At each meeting it receives the Report of the International Committee on the work accomplished, and it is responsible for:

—discussing and instigating the arrangements required to ensure the propagation and improvement of the International System of Units (SI), which is the modern form of the metric system;
—confirming the results of new fundamental metrological determinations and the various scientific resolutions of international scope;
—adopting the important decisions concerning the organization and development of BIPM.

The International Committee consists of eighteen members each belonging to a different State; it meets at least once every two years. The officers of this Committee issue an *Annual Report* on the administrative and financial position of BIPM to the Governments of the Member States of the Metre Convention.

The activities of BIPM, which in the beginning were limited to the measurements of length and mass and to metrological studies in relation to these quantities, have been extended to standards of measurement for electricity (1927), photometry (1937) and ionizing radiations (1960). To this end the original laboratories, built in 1876–1878, were enlarged in 1929 and two new buildings were constructed in 1963–1964 for the ionizing radiation laboratories. Some thirty physicists or technicians work in the laboratories of BIPM. They do metrological research, and also undertake measurement and certification of material standards of the above-mentioned quantities. BIPM's annual budget is of the order of 3 000 000 gold francs, approximately 1 000 000 U.S. dollars.

In view of the extension of the work entrusted to BIPM, CIPM has set up since 1927, under the name of *Consultative Committees*, bodies designed to provide it with information on matters which it refers to them for study and advice. These Consultative Committees, which may form temporary or permanent "Working Groups" to study special subjects, are responsible for co-ordinating the international work carried out in their respective fields and proposing recommendations concerning the amendments to be made to the definitions and values of units. In order to ensure worldwide uniformity in units of measurement, the International Committee accordingly acts directly or submits proposals for sanction by the General Conference.

The Consultative Committees have common regulations (*Procès-*

Verbaux CIPM, 1963, 31, 97). Each Consultative Committee, the chairman of which is normally a member of CIPM, is composed of a delegate from each of the large Metrology Laboratories and specialized Institutes, a list of which is drawn up by CIPM, as well as individual members also appointed by CIPM and one representative of BIPM. These Committees hold their meetings at irregular intervals; at present there are seven of them in existence.

1. *The Consultative Committee for Electricity* (C.C.E.), set up in 1927.
2. *The Consultative Committee for Photometry and Radiometry* (C.C.P.R.), new name given in 1971 to the Consultative Committee for Photometry set up in 1933; between 1930 and 1933 the preceding Committee (C.C.E.) dealt with matters concerning Photometry.
3. *The Consultative Committee for Thermometry* (C.C.T.), set up in 1937.
4. *The Consultative Committee for the Definition of the Metre* (C.C.D.M.), set up in 1952.
5. *The Consultative Committee for the Definition of the Second* (C.C.D.S.), set up in 1956.
6. *The Consultative Committee for the Standards of Measurement of Ionizing Radiations* (C.C.E.M.R.I.), set up in 1958.

Since 1969 this Consultative Committee has consisted of four sections: Section I (measurement of x and γ rays); Section II (measurement of radionuclides); Section III (neutron measurements); Section IV (α-energy standards).

7. *The Consultative Committee for Units* (C.C.U.), set up in 1964.

The proceedings of the General Conference, the International Committee, the Consultative Committees, and the International Bureau are published under the auspices of the latter in the following series:

—*Comptes rendus des séances de la Conférence Générale des Poids et Mesures;*
—*Procès-Verbaux des séances du Comité International des Poids et Mesures;*
—*Sessions des Comités Consultatifs;*
—*Recueil de Travaux du Bureau International des Poids et Mesures* (this compilation brings together articles published in scientific and technical journals and books, as well as certain work published in the form of duplicated reports).

From time to time BIPM publishes a report on the development of the

Metric System throughout the world, entitled *Les récents progrès du Système Métrique*.

The collection of the *Travaux et Memoires du Bureau International des Poids et Mesures* (22 volumes published between 1881 and 1966) ceased in 1966 by a decision of CIPM.

Since 1965 the international journal *Metrologia*, edited under the auspices of CIPM, has published articles on the more important work on scientific metrology carried out throughout the world, on the improvement in measuring methods and standards, of units, etc., as well as reports concerning the activities, decisions and recommendations of the various bodies created under the Metre Convention.

Appendix II

TREATY OF THE METER

The following is an updated abstract from a report entitled Measuring Systems and Standards Organizations, *written by William K. Burton. The entire report (44 pages) is available from the American National Standards Institute, 1430 Broadway, N.Y., N.Y. 10018.*

While Americans reflect on the past 200 years of this nation's growth from revolution and reform to international recognition, perhaps it would be an appropriate time to remember another revolution that was occurring at about the same time—the reform of a customary system of measurement to a new, internationally understood metric system.

May 20, 1975 marks the one-hundredth anniversary of the beginning of international acceptance of SI, the modern metric system of weights and measures. The signing of the Treaty of the Meter by 17 nations, including the U.S., in Paris on May 20, 1875 was a major step in developing a universally acceptable language of measurement.

Birth and Development of the New System

Scientific advances made during the seventeenth and eighteenth centuries provided the initial stimuli for the development of the metric system. During those centuries many new scientific instruments were invented and older ones radically improved.

Scientific leaders like Newton and, later, the chemist Lavoisier needed increasingly accurate methods of measuring to carry on their investigations. They found themselves hampered by a lack of true international standards of measurement in communicating their discoveries to the world at large.

As early as 1585, the Flemish mathematician Simon Stevin had proposed the idea of a decimal system of measurement units. By 1670 Gabriel Mouton, vicar of St. Paul's Church in Lyons, France, was studying pendulum motion and measurement of the arc of a terrestrial meridian as bases for a natural standard of linear measurement.

These ideas continued to attract attention and discussion in the scientific circles of the eighteenth century. In 1722 Cassini advanced the idea of a new international division of a terrestrial meridian. With this idea in mind, he and Lacaille, in 1739–40, measured the arc of the Dunkirk-Barcelona meridian passing through Paris.

French Revolution Real Catalyst

The real catalyst for the development of the metric system was provided by the tremendous social and political upheaval of the French Revolution. In their enthusiasm for the disruption of the most ancient and revered of European traditions, little was more obvious to the men who led the Revolution than a need for radical reformation of existing measurement systems.

In 1790 the National Assembly, prodded by the beliefs of Maurice de Talleyrand, bishop of Autun, began to seriously entertain ideas that led to the establishment of the metric system. By 1793, after considerable discussion of various proposals, a decision had been made to name the new unit of linear measurement the *meter* (one ten-millionth of the meridian quadrant). A provisional value was adopted on the basis of the Cassini-Lacaille measurement.

Kilogram Defined

During this time Lavoisier made experiments to determine the weight in a vacuum of one cubic decimeter of distilled water at the freezing point. This became the basis for a provisional *kilogram*. Although the revolutionary government sent Lavoisier to the guillotine, other scientists were engaged to continue his experiments on a kilogram basis by weighing water at various temperatures, i.e., different densities.

The astronomers Mechain and Delambre were retained to remeasure the Dunkirk-Barcelona meridian with greater precision, a task which would take them many years to complete.

In 1799 the provisional meter and kilogram were abolished and replaced by new, more accurate standards. A platinum meter and kilogram were constructed and placed on deposit in the Institute National des Sciences et des Arts. Copies were made and used as French national standards.

Decrees of 1800, 1812

The primary problem with the new system was the lack of secondary standards, which were never constructed owing to lack of government funds. In addition private individuals often ignored official weights and measures regulations for their own. Both the private groups and officials employed the old as well as the new legal measures, and serious abuses and frauds resulted. Because of the confusion and lack of acceptance, in November 1800 a decree was issued, which stated that the decimal system of weights and measures would definitely be put into operation by the entire republic. In order to facilitate its use, French names were given to the units.

While this decree tended to weaken the integrity of the system, it preserved its fundamental decimal feature. However, a decree by Napoleon in 1812 threatened the existence of this feature and of the metric system itself. Napoleon's *usuelle* system was based on metric but made use of multiples and fractions, which resulted in measures harmonious with those used for commerce in the past.

Twenty-five years later, in 1837, the French government agreed that the legislation of 1812 had been a mistake. A law was then passed making the metric system, once again, the sole system of measurement in France.

USA: Metric Legalized in 1866

While France was developing the metric system in 1790, the United States was busy drawing up its Constitution, which included a provision for fixing standard weights and measures. The new republic decided, however, to retain the English/customary measurement system, since the bulk of its trade was carried on with England. However, the metric system was made legal (but not mandatory) in 1866.

Other acts were passed by Congress; these provided each state with a set of standard metric weights and measures and provided for the distribution of metric balances to all post offices exchanging mail with foreign countries. These acts opened the door for the United States to participate in the international metric events taking place at that time, to eventually (in 1893) accept the international standards of measurement that would be based on metric, and finally to call for a U.S. Department of Commerce study of the effect of a conversion to metric on the United States.

International Acceptance of Metric: Standards Established

With the metric system gradually becoming accepted in France and throughout Europe, the question of the accuracy of its base units became a matter of importance. In 1869 the French government invited various nations to send delegates to a conference in Paris to discuss the construction of a new prototype meter as well as the establishment of a number of identical standards for the participating nations.

This international commission met in 1869, but the proceedings were interrupted by a war between France and Germany. The commission met again in 1872 and was divided into 11 committees. The new standards were to represent the length of a meter at zero degrees centigrade. The meter bar itself was to consist of an alloy of platinum (90%) and iridium (10%). The international kilogram was to be determined with reference to its weight in a vacuum.

In addition to other recommendations, the commission advocated the founding of an international bureau of weights and measures to be located in Paris. This body would be international and neutral, and was to be supported by contributions from each participating nation.

This bureau would be responsible to the General Conference of Weights and Measures (CGPM), meeting every six years, and to the International Committee, meeting every two years.

Treaty of the Meter

On May 20, 1875 a treaty was concluded in Paris wherein the recommendations of the Commission were put into effect: reformulation of the metric system; construction of new measurement standards and distribution of accurate copies to participating countries; permanent machinery for further international action on weights and measures; and a world repository and laboratory (the International Bureau of Weights and Measures near Paris). The treaty was signed by the United States, Germany, Austria-Hungary, Belgium, Brazil, Argentine Confederation, Denmark, Spain, France, Italy, Peru, Portugal, Russia, Sweden, Norway, Switzerland, Turkey, Venezuela, and later, Great Britain.

The new measurement standards, including meter bars and kilogram weights, were finished in 1889, and the U.S. received its copies. Four years later, the U.S. Secretary of the Treasury, by administrative order, declared the new metric standards to be the nation's "fundamental standards" of length and mass. Thus the U.S. *officially* became a metric nation. The yard, the pound, and other customary units are defined as fractions of the standard metric units.

Refinements of the SI System

Over the years two versions of the metric system had been used in science and engineering. The *cgs* system consisted of the centimeter, gram, and second as Base Units, while the *mks* system contained the meter, kilogram, and second. By 1900 measurements in the metric system began to be based on the *mks* system. Later Professor Giorgi of Italy recommended that the units of mechanics should be linked with electromagnetic units, and the ampere was added to create the *mksa* (Giorgi) units.

The Tenth General Conference of Weights and Measures, meeting in 1954, adopted a rationalized and coherent system of units based on the four *mksa* units plus the *kelvin* as the unit of temperature and the *candela* as the unit of luminous intensity. (In 1971 the seventh base unit, the *mole*, was accepted as the base unit of amount of substance.)

In 1960 the Eleventh General Conference of Weights and Measures established the *Système Internationale d'Unités* (SI) as the official title of the internationally accepted metric system.

Since later measurements in the quadrant found that the measurement from which the meter was derived was in error by one part in 3000, a new unvarying and indestructible unit had to be discovered. In 1961 the meter was established as equal to 1 650 763.73 wavelengths of the orange-red light given off by krypton 86. This new determination of the meter has the advantage of being reproducible in scientific laboratories throughout the world.

Since then the CGPM has also changed the means of measuring the second from 1/86 400 of a mean solar day (the average based on the varying revolutions of the earth) to a "duration of 9 192 631 770 periods of the radiation between 2 levels of the cesium 133 atom."

The present SI system is by no means final and unchangeable. It will always be reviewed, subject to change as the CGPM sees fit in its periodic reviews.

The success of the rapid development and acceptance of the metric system can probably be tied to the revolutionary zeal of the newborn French Republic, which was discarding traditional concepts in many areas. The new system was designed around the following principles advocated by scientists and mathematicians:

- A society using a decimal numbering system should have a decimalized measuring system.

- Units of length, volume, and weight should have a direct relationship, one to the other.

- The value of the basic unit of length should be taken from some unchanging, absolute standard found in the physical universe.

Conformance to these principles and the resulting simplicity and ease of application gave the metric system the advantages that led to its international acceptance. It was caught up in the tidal wave of scientific and industrial growth that had its beginnings at the same period. Increasing world trade and technological advances provided the catalyst for its adoption by more and more countries, until at the present time metric countries account for over 90% of the world's gross national product, leaving the United States as the only major industrial nation not using metric.

The opportunities for improvement in international trade and standards, the reduction of the number of sizes, and the simplicity of a decimal-based system have made the metric system attractive to the major industrial countries of the world. However, the economic and physical difficulties of adopting metric units in a highly industrialized nation must be considered as well. Careful study and meticulous planning must accompany national metric conversion plans.

Base SI Units

Unit	Name	Symbol
Length	meter	m
Mass	kilogram	kg
Time	second	s
Electric current	ampere	A
Temperature	kelvin	K
Luminous intensity	candela	cd
Amount of substance	mole	mol

Index

Abandoned units, in non-SI, 21
Absolute system, 32
Absolute versus gravitational system, 32
Acceleration, definition of, 57
Ambiguity, mass-force, 35
Ampere, definition of, 226, 228
Ampere, watt, and joule, relationship between, 119
Angular impulse and momentum, 84, 85
Angular measure in SI, 19
  in rotating machinery, 121–123
  in statics, 34
Area in SI, 12

Bar and the pascal, 50, 51
Base units, 7, 234
Beams, 99
  analysis in SI units, 101–104
  deflections in, 105
  fiber stress in, 100
Bearings, lubrication of, 178–180
Bearing stress, 90
Bending and torsion, in shafts, 126, 127
  in SI units, 18
Bending, direct stress with, 114, 115
Bending moment, 41
Bernoulli's Theorem, 159
BIPM, 4, 289, 290
Buckling of columns, 111–114
Building code changeover to SI, 286, 287
Buoyancy, 149, 150

Calculations, pipe flow, 174–177
  stress and strain, 91–95
Center of pressure, 151
Centrifugal action, circular motion, 76–81
CGPM, 14, 289
Changeover of existing charts, customary to SI, 243–246
Changeover to a coherent system, 264
Changeover to SI, "soft" versus "hard," 26, 27
Changeover versus conversion to SI, 10, 11
Channels, flow in, 188
Choice of unit multiples, 98
CIPM, 4, 289–291
Circular motion, centrifugal action in, 76–81
Coherent system, conversion to a, 264
Column action, 111–114
Comparison of SI with customary practice in fluid mechanics, 180–183
Compatible unit multiples, 96, 97

Composite elastic elements, 109–111
Compound units—single prefixes, 41–44
Compression, 90
Conduction, thermal, 217, 218
Conduits, flow in, 188
Constants, classes of, 270–275
Constants in SI terms, frequently used, 275–278
Construction modules and sizes, 285, 286
Convection, thermal, 218–220
Conversion factors, numerical, 230–231, 237–240, 245, 246, 250, 254–263
Conversion, graphical, 240, 241, 243, 244
  to preferred SI usage, 233–264
  versus changeover to SI, 10, 11
Counterbalancing, 138–140
Customary practice versus SI in fluid mechanics, 180–183
Customary to SI changeover of existing charts, 243–246

Data, precision of, 246–250
Dates, ISO recommendation, 21
Decimalization in SI, 9, 24, 266, 267
Deflections, beam, 105
Density, 141, 279
Derived units, 7, 194–196, 235–237
Development of the SI system, 293, 294
Direct stress with bending, 114, 115
Distance or length in SI, 11, 12
Division sign in SI, 24
Drafting policy, 282, 283
Drafting scales, 283, 284
Dryness fraction, 201–209
Dual-scale charts, single-line, 241–243
Dynamics, 57–87

Elastic elements, composite, 109–111
Elasticity, definition of modulus of, 88
Elasticity, stress, and strain, 88–95
Elastic modulus in SI, 18, 19
Electricity, magnetism, and light, 225–232
Electric units, definitions of, 228
Empirical factors, SI units of, 69
Energy and work, 70
Energy equation, 159–162
  steady-flow, 196–199
Energy, in SI, 19, 20
  kinetic, 71–74
  of rotation, 85–87
  potential, 71, 151–153
  strain, 71

300 INDEX

Energy system, total, 163, 164
Engineer's "*g*", 36
Entropy, 209–213
Equivalents, 267
  of SI quantities, 22, 23
Exponent interpretation in SI, 26

Fiber stress in beams, 100
Flow in pipe, 171, 172, 174–177
Flow meter, 166, 167
Fluid flow, incompressible, 158
Fluid mechanics, 141–189
Flywheels, 130, 131
Force definitions, 15, 30
Force determination, gravitational, 37
Force in dynamics, 61
Force in SI, 16–18
  the direct approach to, 36
Force-mass examples, 44–49
Force of a jet, 167, 168
Force of gravity, 34
Forces, effect of unbalanced, 65, 66
  understanding, 33
Free-falling bodies, 64, 65
Friction factor in fluid flow, 173, 174
Friction head, 164–166
Fuels, heating value of, 224

Gases, properties of, 213–217
Gears, 127–130
General Conference on Weights and Measures (CGPM), 1, 4, 289
"*g*", engineer's, 36
Governors, 136
Graphical conversion, 240, 241
Gravimetric analysis, combustion, 216
Gravitational attraction, 275
Gravitational force determination, 37
Gravitational versus absolute system, 32
Gravity and mass in dynamics, 60
Gravity, force of, 34
  specific, 143
Grouping of numbers in SI, 24
Guidelines for use of SI, 11–21
Gyroscopes, 132

Handling SI units, 105–109
Harmonic motion, simple, 67–69
Head, fluid, 150, 151
  friction loss, 164–166
Heat, 194
Heating value of fuels, 224
Heat transfer and thermodynamics, 190–224
  derived units for, 194–196
Heat transfer, modes of, 217
  overall, 221–224

Impulse and momentum, 66, 67
  angular, 84, 85
Incompressible fluid flow, 158

Inertia and torque, 81–84
International acceptance of metric, established standards, 296
International Bureau of Weights and Measures (BIPM), 4, 289, 290
International Committee for Weights and Measures (CIPM), 4, 289–291

Jet, force of a, 167, 168
Joule, ampere, and watt, relationship between, 119

Kilogram defined, 294
Kinetic energy, 71–74

Lateral strain, 115
Legalization of metric in 1866 in USA, 295
Length or distance in SI, 11, 12
Light, 229–232
Limiting torque, 133, 134
Load, definition of, 32
Losses, pipeline, 177, 178
Lubrication of bearings, 178–180

Machinery, angular measure in rotating, 121–123
Machines, mechanics of, 119–140
  power ratings of, 119
Machining power requirements, 136–138
Magnetism, light, and electricity, 225–232
Manning's equation in SI, 273
Manometers, 153–156
Map, road, 287
Mass and gravity in dynamics, 60
Mass, ascertaining, 44
  definition of, 30
Mass-force examples, 44–49
Mass in SI, 13, 14, 30
Mass versus weight ambiguity, 35
Materials, physical properties of, 92–93
Measuring the rate-of-flow, fluids, 159
Mechanical units which enter the definitions of electric units, definitions of, 227
Mechanics of machines, 119–140
Meter, flow, 166, 167
  venturi, 166, 167
Metre, Treaty of the, 293, 296
Metric legalization in 1866 in USA, 295
Metric module demonstration and SI file, 28, 29
Metric paper sizes, 281, 282
Metric standards, international acceptance of, 296
Metric system, decimal nature of, 266, 267
Mixing of units improper in SI, 26
Modules in SI, 27, 28, 285
Modulus of elasticity, definition of, 88
Modulus of rigidity, 89
Modulus, Young's, 89
Mole, definition of, 214

INDEX

Moment, bending and torsional, 41
Moment in SI, 18
Moment of inertia ($\bar{I}$) in SI, 19
Momentum and impulse, 66, 67
   angular, 84, 85
Motion, definition of, 57
   rectilinear, 61-64
   rotational, 75, 76
   simple harmonic, 67-69
Multiples, choice of unit, 98
   compatible unit, 96, 97
Multiplication sign in SI, 24

NBS unit tables for SI, 233
Newton's laws, 34
Notation and numerical style in SI, 24-26
Numbers, rounding of, 253-260
Numerical conversion factors, 237-240
Numerical style and notation in SI, 24-26

Optimization of sizes in SI, 27
Organizations for SI standards, 6
Orifices, 188

Paper sizes in metric, 281, 282
Pascal and the bar, 50, 51
Pendulum, simple, 69, 70
Physical properties of materials, 92-93
Pipe flow, 171, 172
Pipe flow calculations, 174-177
Pipeline losses, 177, 178
Pitot tube, 168, 169
Potential energy, 71, 151-153
Power in SI, 19, 20, 75
Power ratings of machines, 119
Power requirements, machining, 136-138
Prandtl number, 219
Precision of data, 246-250
Preferred sizes, in SI, 27, 28
   series of, 279-281
Preferred usage, deciding on, 278, 279
Prefixes and compound units, 41-44
Prefixes in SI, 9, 25, 268-269
Pressure, 49, 50, 145-149, 190, 193, 276-277
   center of, 151
   in SI, 18, 19
Pump and turbine characteristics, 184, 185

Quality of vapor, thermodynamics, 201-209

Radiation, heat transfer, 220, 221
Rate-of-flow, measuring the, 159
Ratings of machines, power, 119
Rectilinear motion, 61-64
Refrigeration cycle, 210
Reynolds number, 172, 173
Rigidity, modulus of, 89
Road map, 287
Rotational motion, 75, 76
Rotation, energy of, 85-87
Rounding of numbers, 253-260

Scales, drafting, 283, 284
Second moment ($\bar{I}$) in SI, 19
Second moment of area and section modulus, 100
Section modulus (S) in SI, 19
Sections, properties of structural, 101
Shafting, power, 123-126
Shafts, bending and torsion in, 126, 127
   torsion in, 96
   whirling, 134-136
Shear and moment diagrams, 99
Shear stress, 91
SI, advantages of going, 2, 3
   angular measure in, 19
   area in, 12
   bending and torsion in, 18
   building code changeover to, 286, 287
   comprehending the system, 233
   conversion to, 233-264
   decimalization in, 9
   decimals in, 24
   development of, 293, 294
   direct approach to force in, 36
   division sign in, 24
   elastic modulus in, 18, 19
   energy in, 19, 20
   force in, 16-18
   grouping of numbers in, 24
   guidelines for use of, 11-21
   impact on engineering, 3-6
   introduction to, 1-29
   language of, 3
   length or distance in, 11, 12
   mass in, 13, 14
   merits of, 6
   mixing of units in, 26
   modules in, 27, 28
   moment in, 18
   moment of inertia ($\bar{I}$) in, 19
   moving into the world of, 265-288
   multiplication sign in, 24
   notation and numerical style in, 24-26
   optimization in, 27
   power in, 19, 20
   preferred sizes in, 27, 28
   preferred usage of, 11
   prefixes in, 25
   pressure in, 18, 19
   rationalization in, 27
   reproducibility of units, 10
   second moment ($\bar{I}$) in, 19
   section modulus (S) in, 19
   significant digits in, 21, 22
   "soft" versus "hard" changeover to, 26, 27
   spacing numbers and units in, 24
   standards for, 5, 6
   stress in, 18, 19
   symbolization in, 8, 9
   temperature in, 20
   time in, 20, 21

INDEX

SI, uniqueness of units, 7
   units abandoned in, 21
   universality, 10
   use of exponents in, 26
   volume in, 12, 13
   weight in, 14, 15
SI base units, 7
SI changeover versus conversion, 10, 11
SI derived units, 7
SI file, metric module demonstration and, 28, 29
Significant digits, 251–253
   in SI, 21, 22
Simple harmonic motion, 67–69
Simple pendulum, 69, 70
Simple stresses, 89
Simplification possible with use of SI, 38–41
Single-line, dual-scale charts, 241–243
SI procedure for working examples, 37
SI quantities, equivalents of, 22, 23
SI simplifies computation, examples, 38–41
SI system, refinements of the, 297, 298
SI values, frequently used constants, 275–278
SI versus customary practice in fluid mechanics, 180–183
SI unit charts, 234–237
SI units in beam analysis, 101–104
SI units of empirical factors, 69
SI unit tables, NBS, 233
Sizes of familiar things, 270
"Soft" versus "hard" changeover to SI, 26, 27
Spacing numbers and units in SI, 24
Specific gravity, 143
Specific weight, 143
Springs, 117, 118
Standards, international acceptance of metric, 296
Standards organizations, 6
Statics, 30–56
   angular measure in, 34
   definition of quantities in, 30–32
   examples in, 51–56
Steady-flow energy equation, 196–199
Steam tables, 200, 202, 203
   use of, 199–201
Strain, and stress calculations, 91–95
   definition of unit, 88
   elasticity and stress, 88–95
   lateral, 115, 116
Strain energy, 71
Strength of materials, 88–118
Stress, and strain calculations, 91–95
   bearing, 90
   definition of unit, 88
   shear, 91
   simple, 89
   strain and elasticity, 88–95
   thermal, 91

Stress in beams, fiber, 100
Stress in SI, 18, 19
Stress with bending, direct, 114, 115
Stretch modulus, 89
Structural sections, properties of, 101, 102, 103
Surface tension, 157
Surveying, 284, 285
Symbolization in SI, 8, 9

Temperature in SI, 20, 190
Tensile stress, 90
Tension, surface, 157
Thermal, conduction, 217
   convection, 218
   radiation, 220
   resistance, 222
   stress, 91
Thermodynamics and heat transfer, 190–224
   derived units for, 194–196
Time in SI, 20, 21
Torque and inertia, 81–84
Torque, limiting, 133, 134
Torsional moment, 41
Torsion and bending, in shafts, 96, 126, 127
   units in SI, 18
Total energy system, 163, 164
Treaty of the Metre, 293, 296
Turbine and pump characteristics, 184, 185

Unbalanced forces, effect of, 65, 66
Unit charts, SI, 234–237
Unit multiples, choice of, 98
   compatible, 96, 97
Units, abandoned, 21
   handling, 105–109
Units for thermodynamics and heat transfer, derived, 194–196
Unit strain, definition of, 88
Unit stress, definition of, 88
Universal gas constant, 214

Velocity, definition of, 57
Venturi meter, 166, 167
Vibrations, 132, 133
Viscosity, 169–171
Volume in SI, 12, 13

Watt, joule, and ampere, relationship between, 119
Weight, definition of, 31
   in SI, 14, 15
   specific, 143
Weight versus mass ambiguity, 35
Weirs, 188
Work and energy, 70

Young's modulus, 89

DATE DUE